Vector Calculus

For College Students

Alice Gorguis

Mathematics Department, North Park University,
Chicago, IL. 60625

2nd Edition

To order additional copies of this book, contact:
Xlibris LLC
1-888-795-4274
www.Xlibris.com
Orders@Xlibris.com
646740

Dedication:

This book is dedicated to the memory of my beloved parents who, spiritually, are always with me.

> If we die with Him, we will also live with Him.
> If we endure hardship, we will reign with Him.
> If we are faithful, He remains faithful,
> for He can not deny Himself.
>
> {Timothy 2:11-13}

Preface:

This second edition differs from the first one in the followings: Additional explanation, and examples are added to some sections. Solutions to the execersise problems are included.

The text is intended for a one-semester course in the Calculus of functions of several variables and vector analysis taught at college level. This course is, normally known as , vector calculus, or multi variable calculus, or simply calculus-III. The course usually is preceded by a beginning course in linear algebra. The prerequisite for this course is the knowledge of the fundamental of one-variable calculus, differentiation and integration of the standard functions. The text includes most of the basic theories as well as many related examples and problems. There are many exercises throughout the text, which in my experience are more than enough for a semester course in this subject. I include enough examples for each topics in each section to illustrate and help the student to practice his/her skills. Also, added problems that ask the student to reflect on and explore in his/her own words some of the important ideas of Vector Calculus.

I have included material enough to be covered during a simple semester without a hassle, and it should be possible to work through the entire book with reasonable care. Most of the exercises are relatively routine computations to moderate and productive problems, to help the students understand the concept of each topic.

Each section in a chapter is concluded with a set of exercises that review and extend the ideas that was introduced in the chapter, or section.

Computer softwares were not included in this book. Most of the exercises can be solved easily by hand, but I advise the students to use Mathematica, or Maple to graph the functions in each problem to visualize the problem, and understand it better. Some of the homework might require the use of Mathematica.

In this text we study the calculus of vector field. In particular we define line integrals, surface integrals, the connection between these integrals and the single, double, and triple integrals which are given by the higher-dimensional analogue of the Fundamental Theorem of Calculus: Greens' Theorem, Stokes Theorem, and the Divergence Theorem.

Alice Gorguis
June 2014

Contents

Chapter 1

Euclidean Space

In space, to study the calculus for functions of one-variable, the domain is a part of one-dimensional line R, and its graph is a set of points in two-dimensional plane R^2. Thus in studying functions of two-variables, higher dimensional space will be used, and their graphs will turn out to be a surface in 3-space, and functions of 3-variables will form a surface in 4-space, and so on.

In this chapter we will start with the basic operations of the vectors in two and three dimensional space: vector addition, scalar multiplication, dot and cross products.

1.1   Introduction:

There are two types of physical quantities in real life: one that has magnitude and direction called **Vector** quantities, such as: force, velocity, acceleration, and displacement of moving particles, and these quantities are represented by a directed line segment, and the length of the segment, to some scale, represents the "Magnitude" of the quantity as the one in fig(1).

11

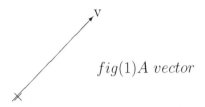

$fig(1) A\ vector$

The direction of this quantity is represented by the inclination of the segment, or the angle between the segment and the horizontal line, or by the arrowhead.

And the other quantity such as area, mass, time, and distance, these can be described by a single number, and these are called **scalar** quantities, and the numbers used to measure them are called **scalars**. For our purposes we will consider scalars as real numbers.

Physical Quantities

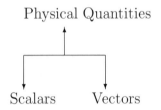

Scalars   Vectors

Historical Notes

Vectors are considered an important tools in physical sciences and Engineering,and their application has extended to the area of Economics and some of the other Social Sciences, also vectors have contributed significantly to the counting development of Mathematical Sciences.

The idea of vector analysis was discovered earlier by the Irish Mathematician William Rowan Hamilton (1805-1865). In his work of quaternions, and the Scottish physicist James Clark

Maxwell(1831-1879), who used some of Hamilton's Ideas in his study of electro Magnetic field Theory. Later in the nineteenth century the American physicist Josiah Willard Gibbs(1839-1903), and the English engineer Oliver Heaviside(1850-1925), developed the subject of vector analysis independently.

1.2   Vector Algebra:

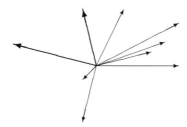

In two dimensional space the point is represented by ordered pairs of real numbers x, and y, and is written as $p(x, y)$ shown in fig(2). Likewise, in three dimensional space the point is represented by three real numbers x, y, and z, and is written as $p(x, y, z)$ shown in fig(3).The line that has its tail at the origin, and its head ends at the point $p(x, y)$ in two dimensions, and $p(x, y, z)$ in three dimensions, is called a vector line. The length of the vector line measured from its tail to its head at the point is called the magnitude, and its direction is measured by the angle θ between the vector and the $x - axis$ as shown in $fig(2)$.

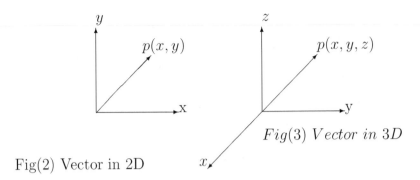

Fig(2) Vector in 2D

Fig(3) Vector in 3D

Definition:

A vector is a line segment with magnitude and direction.

The magnitude of the vector can be represented by the hypotenuse side of a right triangle, hence can be computed using the Pythagorean theorem:

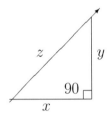

Where, $z = \sqrt{x^2 + y^2}$, and its direction is measured by the angle between the vector and the $x - axis$, in degree measures or radian measures, where one radian $\pi = 180^0$.

Sometimes, we will represent the vector in 2-dimensional space, R^2, as $v = (x, y), or < x, y >$, and the vector in 3-dimensional space,R^3, as $v = (x, y, z), or < x, y, z >$.

Arithmetic Operations with vectors:

Vector Addition:

Definition: Let two vectors with components in 2-dimensional space be given as: $v = (v_1, v_2)$, and $u = (u_1, u_2)$, the vector sum of these two vectors $v + u$ will form also a 2-dimensional vector.

Definition: Let two vectors with components in 3-dimensional space be given as: $v = (v_1, v_2, v_3)$, and $u = (u_1, u_2, u_3)$, the vector sum of these two vectors $v + u$ will form also a 3-dimensional vector.

Example-1.1:
Let vector $u = i + 3j$ and vector $v = 7i + 2j$, be two vectors in $\mathbf{R^2}$, where i, and j are unit vectors along x, and y axis respectively. then the sum of these two vectors:

$$
\begin{align}
u + v &= i + 3j + 7i + 2j \tag{1.1}\\
&= i + 7i + 3j + 2j \tag{1.2}\\
&= 8i + 5j \ in \ \mathbf{R^2} \tag{1.3}
\end{align}
$$

Example- 1.2:
Similarly in $\mathbf{R^3}$, let : Vector $u = -2i + 3j + 2k$, and vector $v = 7i - j + 12k$, then their sum will be:

$$
\begin{align}
u + v &= -2i + 3j + 2k + 7i - j + 12j \tag{1.4}\\
&= -2i + 7i + 3j - j + 2k + 12k \tag{1.5}\\
Then, \ u + v &= 5i + 2j + 14k \ in \ \mathbf{R^3} \tag{1.6}
\end{align}
$$

Homework-1:

(a) Express the vector $v = AB$ in algebraic form, where $A = (-3, 2)$ and $B = (5, -5)$. Draw the geometric vector.

(b) Draw the Geometric representative of the vector $(-1, 4)$ whose initial point is $(3, -2)$. Find its terminal point.

Note: Sometimes, i, j, and k can be just neglected by writing the vectors as follows:The sum in (1.3) can be written as: $u + v = < 8, 5 >$ in $\mathbf{R^2}$, and the sum in (1.6)can be written as: $u + v = < 5, 2, 14 >$ in $\mathbf{R^3}$, where the symbols $< >$ represents vector.

As we can see from the above examples, vectors can be described in 4-different ways in $\mathbf{R^3}$(and $\mathbf{R^2}$) as follows:

1. As (a,b,c), as ordered triples.
2. As (a,b,c), as a point in the space.
3. As a directed line segment with initial point (x_0, y_0, z_0), and terminal point $(x_0 + a, y_0 + j, z_0 + k)$.
4. As $ai + bj + zk$.

Properties of vector Addition:

Vectors like numbers have properties that apply in $\mathbf{R^2}$, and $\mathbf{R^3}$ as follows:

1. Cumulative Property: $u + v = v + u$.
2. Associative Property: $u + (v + w) = (u + v) + w$.
3. Zero Property: $u + 0 = u$

Homework-2: Proof the above 3 Properties.

Homework-3:

A man and a woman are pulling a sled by ropes of 5-feet long attached to the sled at front center. The woman is holding the rope 3-feet above the sled and 3-feet to the side of the point of attachment, and the man holds the rope 4-feet above the sled and 3-feet to the opposite side of the point of the attachment.The woman exerts a force (F_w) of magnitude 6-pounds, and the man a force (F_m) of magnitude of 8-ponds. Find the resultant force $F_w + F_m$ on the sled.

Homework-4:

Two men are pulling a heavy box, fig(4), if one of the men exerts a force of 800 pounds on the cable tied at one side with angle 30. What force must be exerted on the other side of the cable at angle of 45, if the box is to move along the x-axis.

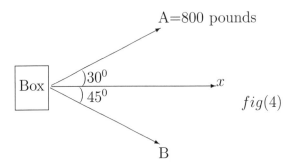

$fig(4)$

Properties of Scalar Multiplications:

Let u and v be vectors in $\mathbf{R^2}$ or $(\mathbf{R^3})$ and let k and l be scalars in $\mathbf{R}$(real numbers), then the following properties hold:

1. Distributive Property: $(k+l)u = ku + lu$.
2. The associative property: $kl(u+v+w) = k[l(u,v,w)]$.
3. Zero property: $k(0,0,0) = (0,0,0)$.
4. Identity property: $I(u,v,w) = (u,v,w)$.

Homework-5: Proof the above 4 Properties.

Geometric Illustration of Vector Addition

1. If two vectors v_1 and v_2 have different magnitude, but same direction, or are collinear, then their sum is simply $v_1 + v_2$ algebraically as shown in fig(5).

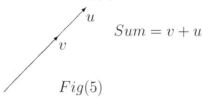

$$Sum = v + u$$

$$Fig(5)$$

2. If two vectors v_1, and v_2 have different magnitudes, and different directions, or not collinear,fig(6), then their sum $v_1 + v_2$ is computed geometrically as follows:

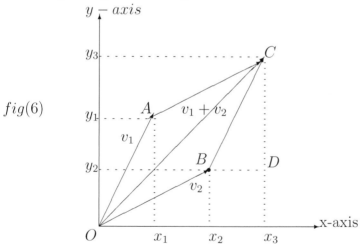

From the graph, $v_1 = (x_1, y_1)$, and $v_2 = (x_2, y_2)$. We need to proof that the sum of the two vectors $v_1 + v_2 = (x_3, y_3)$.

In the parallelogram we note the following triangles: OAX_1 similar to BCD then: the sides, $OA = BC$, $Ox_1 = BD$; Since BDx_3x_2 is a rectangle, then $BD = x_2x_3$. Hence :
$Ox_1 = BD = x_2x_3$. Since : $Ox_3 = Ox_2 + x_2x_3$; then
$Ox_3 = Ox_2 + Ox_1$. In a similar way, on $y - axis$ we can proof that: $Oy_3 = Oy_2 + Oy_1$, which proves that $v_1 + v_2 = (x_2, y_2)$.
3. Another method to illustrate vector addition is using triangles method or (Head to Tail Method) fig(7): we translate the line that represent vector v_2 so that it begins at the end of vector v_1, then connecting the end point with the initial point gives the sum of the two vectors $v_1 + v_2$ and is called the **resultant**.

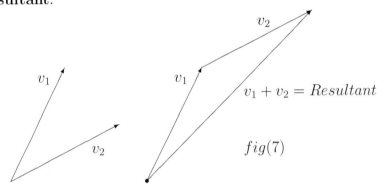

Vectors are positive if pointed on positive direction of x and y axis, and negative if pointed on the negative direction of x and y axis.

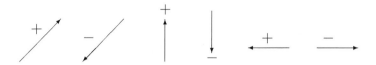

Magnitude of a Vector:

For a vector $u = <a, b>$ the magnitude of u or its length can be measured by using **Pythagorean Theorem**, *where the formula can be written as:*

$$\|u\| = \sqrt{a^2 + b^2} \tag{1.7}$$

A Unit Vector:

A unit vector is a vector with magnitude, or length, equal to one unit. Unit vector is measured by dividing the vector by its scalar length, or magnitude for example vector u its unit vector will be: $\frac{u}{\|u\|}$.

1.3   Exercise -1:

1. Graph the vector in three dimensions $u = i + 4j + 3k$.

2. Graph the following Vectors in $\mathbf{R}^2$
a) (-2,2).
b) (3,2).
c) (3,-2).
d) (-3,-2).

3. Add the two vectors:
u= 1/4(12, 4, 8). v= 2(1, 2, 3).

4. Graph the vectors: u=(2,3), v=(-2,4) and u+v using head to tail method.

5. Graph the vectors u = (2,4,2), v=(0,-4,6), and u+v, and find: -1(2,4,2), 4(0,-4,6).

6. If the sum of two vectors is given as:
(x, 8, 5) + (6, y, -8) = (4, -2, z). Find x, y, and z.

7. Find the magnitude (length) of the vectors:
u=(5,3) in $\mathbf{R}^2$.

8. Find the direction of the vector u= $(1, \sqrt{3})$.

9.Find $\|u\|$ for each of the following vectors:

(a). $u = (-4, 3)$.
(b). $u = -22i + 15j$.
(c). $u = -8i - 3j$

10. If $u = -2i - 5j$, and $v = 3i + 2j$. Find the unit vector for each of the following vectors:
(a). $\|1/2u + v\|$.
(b). $\|2u - v\|$.
(c). $\| - 2u + 3v\|$.

11. Find the unit vector for the vector u in each of the following:
(a). $u = -2i - 3j$.
(b). $u = -5i + 9j$.
(c). $u + 12j + 17j$.

12. Two forces $F_1 = 300$ pounds, and$F_2 = 450$ ponds are acting on an object at rest, from opposite directions along the $x-axis$. Find the resultant force, its magnitude and direction. See solutions on page-163.

1.4   Dot Product (•)

Now that we have defined the sum and difference of two vectors, we need to define the product of two vectors. In product one simple way is to multiply the vectors, component wise, as was done in adding and subtracting vectors , but first we will start with the product of unit vectors to make it easy for the students to understand vector multiplications. We define i, j, k unit vectors as follows:

$$
\begin{aligned}
i &= (1, 0, 0). \\
j &= (0, 1, 0). \\
k &= (0, 0, 1).
\end{aligned}
$$

These vectors satisfies the following relationships:
$i.i = j.j = k.k = 1$, when vectors are parallel to each other, where angle $\theta = 0$.
$i.j = j.k = k.i = 0$, when vectors are perpendicular to each other, where angle $\theta = 90$.

Theorem-1: If vector $u = (u_1, u_2, u_3)$, then $u = u_1 i + u_2 j + u_3 k$. Conversely, if vector $u = u_1 i + u_2 j + u_3 k$, then $u = (u_1, u_2, u_3)$.

Proof:

$$
\begin{aligned}
(u_1, u_2, u_3) &= (u_1, 0, 0) + (0, u_2, 0) + (0, 0, u_3) \\
&= u_1(1, 0, 0) + u_2(0, 1, 0) + u_3(0, 0, 1) \\
&= u_1 i + u_2 j + u_3 k
\end{aligned}
$$

The first type of product vectors that we define is the **dot − product**, sometimes referred to as the **inner − product**, or **scalar − product**, and it is denoted by $u.v$.

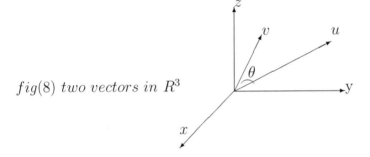

$fig(8)\ two\ vectors\ in\ R^3$

The dot-product is very useful in the physical applications and have interesting geometrical interpretations. Suppose we have two vectors u, and v in R^3 as shown in fig(8), to determine the angle between u and v, the dot-product will help us to do that:Let

$$
\begin{aligned}
u &= u_1 i + u_2 j + u_3 k, \\
v &= v_1 i + v_2 j + v_3 k.
\end{aligned}
$$

$$
Then,\ u.v = u_1 v_1 + u_2 v_2 + u_3 v_3 = scalar.
$$

Where, $i.i = j.j = k.k = 1$, and $i.j = j.k = k.i = 0$. In vector form, the above dot-product is written as $u.v = < u, v >$.

Properties of Dot-Products:

If u, v and w are vectors in R^3, and k, l are real numbers , then:

a) $u.u \geq 0$,

b) $u.u = 0$, Iff $u = 0$.

c) $\alpha u.v = \alpha(u.v)$ and $u.lv = l(u.v)$

d) $u.(v + w) = u.v + u.w$ and $(u + v).w = u.v + v.w$

e) $u.v = v.u$.

Magnitude (length) of the Vector:

The length of vector $u = u_1 i + u_2 j + u_3 k$ can be computed by using Pythagorean formula:

$$u = \sqrt{u_1^2 + u_2^2 + u_3^2} = \|u\|. \tag{1.8}$$

$\|u\|$ in fig(9) is the length of the vector u also known as the norm, where:

$$\|u\| = (u.u)^1/2. \tag{1.9}$$

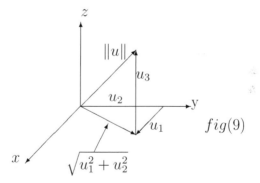

$fig(9)$

Example-1.3:

Find $\|5u - 2v\|$ for the given vectors $u = (3, 2, -2)$, and $v = (-2, -6, 3)$.

Solution:

$$5u - 2v = 3(1, -2, 2) - 2(-3, -4, 5)$$

$$= (3, -6, 6) - (-6, -8, 10)$$
$$= (9, 2, -4).$$

Then :
$$\|3u - 2v\| = \sqrt{81 + 4 + 16} = \sqrt{101}.$$

In plane, to describe the vector u in terms of its components on x and $y - axis$ and the direction or angle θ, we write:
The components of u along the x-axis are:

$$u_x = \|u\|cos\theta. \tag{1.10}$$

And the components of u along the y-axis are:

$$u_y = \|u\|sin\theta. \tag{1.11}$$

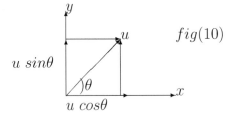

$fig(10)$

In R^2

$$\|u\| = (u.u)^{\frac{1}{2}} \tag{1.12}$$
$$= (cos^2\theta + sin^2\theta)^{\frac{1}{2}}. \tag{1.13}$$

Corollary: The set of unit vectors $\{i, j, k\}$ form orthogonal sets of vectors.

Proof: The set of the given unit vectors can be written as:

$$i = (1, 0, 0),$$
$$j = (0, 1, 0),$$
$$k = (0, 0, 1).$$

Using the dot products we get:

$$i.j = (1, 0, 0).(0, 1, 0) = 0,$$
$$j.k = (0, 1, 0).(0, 0, 1) = 0,$$
$$k.i = (0, 0, 1).(1, 0, 0) = 0.$$

$$\boxed{u.u = \|u\|^2}$$

The Cauchy -Schwartz Inequality:

The dot product satisfies the rule: $u.u = \|u\|^2$, then the size of $a.b$ is measured by, $u.v = \|u\|\|v\|$. This result is known as the Cauchy-Schwartz Inequality.

Theorem::
The inequality $\|u.v\| \leq \|u\|\|v\|$, holds if and only if, one of the vectors is a scalar multiple of the other.

Proof:
If one of the two vectors either u, or v is zero, then both of the inequality are zero, hence the equality holds.
Assuming $u \neq 0$, then we can define a function F for some real number x as:

$$\begin{aligned}F(x) &= \|ux + v\| \\ &= (ux + v).(ux + v) \\ &= \|u\|^2 x^2 + 2u.vx + \|v\|^2\end{aligned}$$

Using the method of completing the square:

$$
\begin{aligned}
F(x) &= x^2\|u\|^2 + 2u.vx + \|v\|^2 \\
&= (x^2\|u\|^2 + 2u.vx) + \|v\|^2 \\
&= \|u\|^2(x^2 + \frac{2u.v}{\|u\|^2}x) + \|v\|^2 \\
&= \|u\|^2(x^2 + \frac{2u.v}{\|u\|^2}x \mp \frac{(u.v)^2}{\|u\|^2}) + \|v\|^2 \\
&= \|u\|^2(x + \frac{u.v}{\|u\|^2})^2 - \frac{(u.v)^2}{\|u\|^2} + \|v\|^2, \\
F(x) &= \|u\|^2(x + \frac{u.v}{\|u\|^2})^2 + \|v\|^2 - \frac{u.v}{\|u\|^2}
\end{aligned}
$$

Hence, $F(x)$ is:

$$
F(x) = \|u\|^2(x + \frac{u.v}{\|u\|^2})^2 + \frac{\|u\|^2\|v\|^2 - (u.v)^2}{\|u\|^2} \tag{1.14}
$$

It is obvious from equation (1.14) that $F(x) \geq 0$, and the minimum value gives the last part of (1.14).

$$
\begin{aligned}
\frac{\|u\|^2\|v\|^2 - (u.v)^2}{\|u\|^2} &\geq 0 \\
\|u\|^2\|v\|^2 - (u.v)^2 &\geq 0 \\
-(u.v)^2 &\geq -\|u\|^2\|v\|^2 \\
(u.v)^2 &\leq \|u\|^2\|v\|^2 \\
\|u.v\| \leq \|u\|\|v\|.
\end{aligned}
$$

Then the equality holds if and only if $F(x) = 0$. $F(x)$ can be zero , IFF, there is a real number x_0 for which: $x_0u + v = 0$, then $v = -x_0u$. Then the Cauchy-Schwartz Inequality has the following interval:

$$
-1 \leq \frac{u.v}{\|u\|\|v\|} \leq 1 \tag{1.15}
$$

From (1.15) it is clear that there is a unique angle θ where, $0 \leq \theta \leq \pi$, for which $cos\theta = \frac{u.v}{\|u\|\|v\|}$. the value of $cos\theta$ is $+1$ on the first quadrant, and -1 on the second quadrant.

Theorem: If u, and v are two vectors in R^2, or R^3, then:

$$u.v = \|u\|\|v\|cos\theta. \tag{1.16}$$

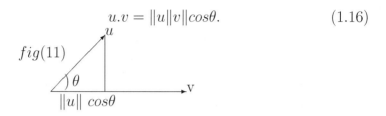

$fig(11)$

Proof: If either u, or v are zero vectors, then :
For $u = 0 \Rightarrow u.v = (0,0,0).(v_1, v_2, v_3) = 0$, and $\|u\| = 0$.
Similarly, if $v = 0 \Rightarrow u.v = (u_1, u_2, u_3).(0,0,0) = 0$, and $\|v\| = 0$.
But if neither u, nor v are zero, then let w be a third vector where $w = v - u$, then applying the law of cosines to the triangle whose sides are u, v, and w gives:

$$\|w\|^2 = \|u\|^2 + \|v\|^2 - 2\|u\|\|v\|cos\theta \tag{1.17}$$

Then from this we get:

$$\begin{aligned} 2\|u\|\|v\|cos\theta &= \|u\|^2 + \|v\|^2 - \|w\|^2 \\ &= u.u + v.v - w.w. \end{aligned}$$

Using the dot-product rules, and $w = v - u$:

$$\begin{aligned} w.w &= (v - u).(v - u), \\ &= (v - u).v - (v - u).u, \\ &= v.v - u.v - v.u + u.u, \end{aligned}$$

Equation (1.17) can be written as:

$$
\begin{aligned}
2\|u\|\|v\|cos\theta &= u.u + v.v - (v.v - u.v - v.u + u.u), \\
&= u.v + v.u, \\
&= 2u.v
\end{aligned}
$$

$$Then, \quad 2\|u\|\|v\|cos\theta = 2u.v$$

And:

$$cos\theta = \frac{2u.v}{2\|u\|\|v\|} = \frac{u.v}{\|u\|\|v\|} \tag{1.18}$$

From (1.18) we get the value of angle θ,

$$\theta = cos^{-1}\{\frac{u.v}{\|u\|\|v\|}\} \tag{1.19}$$

$$\boxed{\theta = cos^{-1}\{\frac{u.v}{\|u\|\|v\|}\}} \qquad Where, \ 0 \leq \theta \leq \pi.$$

Example - 1.4:

Use the above formula to find the angle θ between the two vectors: $u = 2i - j + 2k$, and $v = i - j$.

Solution:

First we need to find the following:

$$
\begin{aligned}
u.v &= (2i - j + 2k).(i - j) \\
&= (2, -1, 2).(1, -1, 0) \\
&= 3
\end{aligned}
$$

And, $\|u\| = \sqrt{2^2 + 1^2 + 2^2} = 3$; $\|v\| = \sqrt{1^2 + 1^2} = \sqrt{2}$

Substituting in the above formula gives:

$$\theta = \cos^{-1}\{\tfrac{3}{3\sqrt{2}}\} = \cos^{-1}(\tfrac{1}{\sqrt{2}}) = 45 = \tfrac{\pi}{4}$$

NOTE: If u, and v are nonzero vectors, then: $\cos\theta = 0$, if and only if $u.v = 0$, and $\cos\theta = 1$, if and only if $u.v = 1$.

Example - 1.5:
Show that the two vectors: $u = (1, 1, 0)$, and $v = (1, -1, 1)$ are perpendicular.
Solution:
If u, and v are perpendicular, then $\theta = 90$, and $cos\theta = 0$, also $u.v = 0$. Then we need to proof that $u.v = 0$:
$u.v = (1, 1, 0).(1, -1, 1) = 1 - 1 + 0 = 0$.

Example - 1.6:
Find x such that the two vectors: $< x + 2, 1 >$, and $< x, 3 >$ are perpendicular to each other.

Solution:

$$Let, \ u \ = \ < x + 2, 1 > .$$
$$And, \ v \ = \ < x, 3 > .$$

If these two vectors are perpendicular, then:

$$
\begin{aligned}
u.v \ &= \ < x + 2, -1 > . < x, 3 > . \\
&= \ x(x + 2) - 1(3). \\
&= \ x^2 + 2x - 3.
\end{aligned}
$$

$$= (x+3)(x-1).$$
$$Then,\ u.v\ =\ (x+3)(x-1).$$

$$Since,\ u.v = 0\ =\ (x+3)(x-1).$$

Then, $x = -3$, and $x = 1$ which gives:
$u = <-1, -1>$, and $v = <-3, 3>$ when $x = -3$.
And, $u = <1, -1>$, $v = <1, 3>$ when $x = 1$.

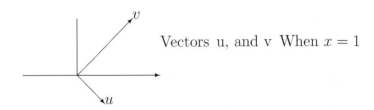

Vectors u, and v When $x = 1$

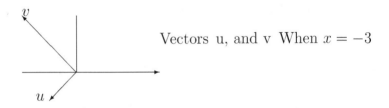

Vectors u, and v When $x = -3$

Dot Product of 2-nonzero vectors with different angles:

For two vectors u, and v with angle θ between them:

1. If $\theta = 0°$, then:

$$\theta = 0$$

$$u.v = \|u\|\|v\|cos0 = \|u\|\|v\|$$

2. If $\theta = 45°$, then:

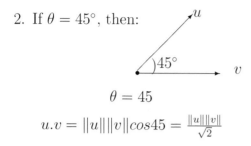

$$\theta = 45$$

$$u.v = \|u\|\|v\|cos45 = \frac{\|u\|\|v\|}{\sqrt{2}}$$

3. If $\theta = 90°$:(right angle)

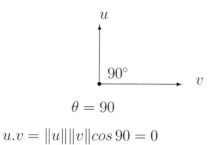

$$\theta = 90$$

$$u.v = \|u\|\|v\|cos\,90 = 0$$

4. If $\theta = 180°$:(straight angle)

$$u.v = \|u\|\|v\|cos\,180 = -\|u\|\|v\|$$

5. If $\theta = 270°$

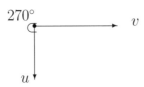

$$u.v = \|u\|\|v\|\cos 270 = 0$$

6. If $\theta = 360°$ (*Circular angle*)

360°

$$u.v = \|u\|\|v\|\cos 360 = \|u\|\|v\|$$

1.5   Exercise - 2

For problems (1 - 4) compute u, v, $\|u\|$, and $\|v\|$:

1) u= (2,3), v=(-3,5).

2) u=(2,-6,-4), v=(1,-2,-5).

3) u = (-2i+3j-k), v=(-1/2 i+3/2j+k).

4) u=(0,0,3), v=(-3,0,-1).

For the problems (5 - 7) find the angle between the two vectors u, and v:

5) $u = i + \sqrt{2}j$, $v = i - \sqrt{2}j$.

6) u= (1,-1,1), v= (-1, -2, -2).

7) $u = i+j+k$, v= -i +2j+3k.

8) Compute the unit vector for the vector $u = (-2, 0, 1)$.

9) Compute the vector u of length 3-units that points on the direction of $(1, 1, -1)$.

10) Find two non parallel vectors that are perpendicular to $(i - j + k)$.

11) For the given vectors: $u = i$, and $v = j$ determine if they are perpendicular.

12) For the given vectors: $u = \sqrt{2}i - 3j + 2k$, and $v = i + \sqrt{2}j + 3k$ determine if they are perpendicular.

13) Compute $u.v$ where: $u = \sqrt{2}i - 30j + 15k$, and $v = \frac{u}{\|u\|}$.
See solutions on page-166.

1.6   Cross-Product $(\times)$

The second type of vector multiplication is called the cross-multiplication. As we have seen before for the dot-product, the product of two vectors gives a scalar. But for the cross-product the product of two vectors gives a vector too.

The cross-product is a product defined only for vectors in R^3.
Definition: The cross-product of two vectors u, *and* v with ,

$$u = (u_1 i + u_2 j + u_3 k), \; and,$$
$$v = (v_1 i + v_2 j + v_3 k),$$

is a vector denoted by,

$$u \times v = (u_1 i + u_2 J + u_3 k) \times (v_1 i + v_2 j + v_3 k)$$

$$= \begin{bmatrix} i & j & k \\ u_1 & u_2 & u_3 \\ v_1 & v_2 & v_3 \end{bmatrix}$$

$$= i \begin{bmatrix} u_2 & u_3 \\ v_2 & v_3 \end{bmatrix} - j \begin{bmatrix} u_1 & u_3 \\ v_1 & v_3 \end{bmatrix} + k \begin{bmatrix} u_1 & u_2 \\ v_1 & v_2 \end{bmatrix}$$

$$= i(u_2 v_3 - u_3 v_2) + j(u_3 v_1 - u_1 v_3) + k(u_1 v_2 - u_2 v_1)$$

Example- 1.7: Find $u \times v$ if $u = (1, 2, 3)$ and $v = (-2, 1, 4)$.

Solution:

$$u \times v = (1, 2, 3) \times (-2, 1, 4)$$

$$= \begin{bmatrix} i & j & k \\ 1 & 2 & 3 \\ -2 & 1 & 4 \end{bmatrix}$$

$$= i(2(4) - 3(0)) - j(1(4)) - 3(-2))n + k((1(1) - 2(-2))$$
$$= 5i - 10j + 5k$$
$$= 5(1, -2, 1)$$

Properties of Cross Product:

$$1.\ u \times v = -v \times u. \tag{1.20}$$

$$2.\ u \times (v + w) = u \times v + u \times w. \tag{1.21}$$

$$3.\ u \times (av) = ((au) \times v = a(u \times v). \tag{1.22}$$

$$4.\ u \times u = 0 \tag{1.23}$$

For a set of unit vectors i, j, k the cross-product applies as follows:

$$5.\ i \times i = j \times j = k \times k = 0.$$

$$6.\ i \times j = k, j \times k = i, k \times i = j.$$

$$7.\ j \times i = -k, k \times j = -i, i \times k = -j.$$

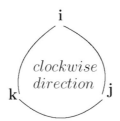

Proof: Left for the students as a homework.

Corollary:

The vector u x v is orthogonal to both u, and v.

Proof: From the rules of dot-product we know that : two vectors can be orthogonal to each other, IFF, their dot-product is equal zero:

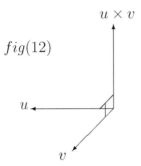

$fig(12)$

$$u.(u \times v) = \begin{bmatrix} u_1 & u_2 & u_3 \\ u_1 & u_2 & u_3 \\ v_1 & v_2 & v_3 \end{bmatrix}$$
$$= u_1 u_2 v_3 - u_1 u_3 v_2 - u_2 u_1 v_3 + u_2 u_3 v_1 + u_3 u_1 v_2$$
$$- u_3 u_2 v_1 = 0.$$

Theorem: The cross-product of two, nonzero, vectors with angle θ between them, can be written as:

$$\|u \times v\| = \|u\| \|v\| sin\theta. \tag{1.24}$$

Proof: we will start with squaring the left side of equation(1.24) as follows:

$$\|u \times v\|^2 = (u_2 v_3 - u_3 v_2)^2 + (u_3 v_1 - u_1 v_3)^2 + (u_1 v_2 - u_2 v_1)^2.$$
$$= (u_2 v_3)^2 + (u_3 v_2)^2 - 2u_2 v_3 u_3 v_2 + (u_1 v_3)^2 + (u_2 v_1)^2$$
$$- 2u_1 v_3 u_3 v_1 = \|u\|^2 \|v\|^2 - (u.v)^2.$$

But from the dot-product rule : $u.v = \|\|u\|\|v\|\cos\theta$.

$$
\begin{aligned}
Then, \quad \|u \times v\|^2 &= \|u\|^2\|v\|^2 - (\|u\|\|v\|\cos\theta)^2. \\
&= \|u\|^2\|v\|^2 - \|u\|^2\|v\|^2\cos^2\theta \\
&= \|u\|^2\|v\|^2(1 - \cos^2\theta) \\
&= \|u\|^2\|v\|^2\sin^2\theta
\end{aligned}
$$

From the last step we can see that: $\|u \times v\| = \|u\|\|v\|\sin\theta$.

The Geometric Interpretation of Cross-Product

The cross -product can be used to find the area of parallelogram, and the volume of a parallelepiped, where:

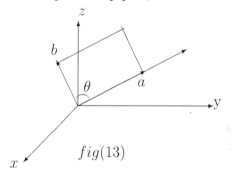

$$fig(13)$$

1. The area of a parallelogram with sides u, and $v = \|u \times v\|$. Where, $\|u\|$ represent the length of the base of the parallelogram, and the altitude has a length of $\|v\|\sin\theta$. Consequently the area of the parallelogram is:

$$\|u\|\|v\|\sin\theta = \|u \times v\|. \qquad (1.25)$$

Example-1.8:
For the given adjacent vectors: $u = i - 2j + 3k$, and
$v = -3i + j + k$. Find the area of the parallelogram inclosed.

Solution:
Using the formula : Area of parallelogram $= \|u \times v\|$ gives.

$$
\begin{aligned}
u \times v &= (1, -2, 3) \times (-3, 1, 1) \\
&= \begin{bmatrix} i & j & k \\ 1 & -2 & 3 \\ -3 & 1 & 1 \end{bmatrix} \\
&= i(-2 - 3) - j(1 + 3(3)) + k((1 - 6) \\
&= -5i - 10j - 5k
\end{aligned}
$$

Then the area of the parallelogram is:

$$
\begin{aligned}
\|u \times v\| &= \sqrt{(-5)^2 + (-10)^2 + (-5)^2}. & (1.26) \\
&= \sqrt{150} \simeq 12.25. & (1.27)
\end{aligned}
$$

2. The area of a triangle with sides u, and $v = 1/2 \, \|u \times v\|$.
If we are looking for the area of a triangle with two sides u,
and v, then we take $1/2$ of the above area in equation (1.27)
or area $\simeq 6.12$.

Example- 1.9:
Find the area of the triangle with the given vertices:

$$
\begin{aligned}
A &= 2i + j + 4k. \\
B &= 3i - j + 7k. \\
C &= -i + 2j + 5k.
\end{aligned}
$$

Solution:
Using the formula for the triangle area we need to find the vectors u, and v first. Let vector,

$$
\begin{aligned}
u &= AB = B - A \\
  &= (1, -2, 3), \\
v &= AC = C - A \\
  &= (-3, 1, 1).
\end{aligned}
$$

Then the area of the triangle is:
$1/2 \, \|u \times v\| = 1/2 \, \sqrt{150} = 5/2 \, \sqrt{6}$.

3. The volume of a parallelepiped with sides u, v, and $w = \|u.(v \times w)\|$.

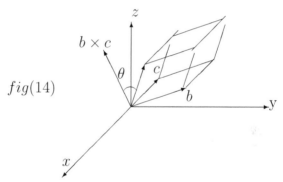

$fig(14)$

Since the volume of parallelepiped is measured by multiplying the area of the base times the hight, and the area of the base is $= \|u \times v\|$, and the hight is $= \|u\| \|v\| \|cos\theta\|$, then,
The volume of parallelepiped $= \|u\| \|u \times v\| \|cos\theta\| = u.(v \times w)$.

Example- 1.10:
Given the following , non-co planner, points:

$$A = (5, 1, 3).$$
$$B = (4, 5, 0).$$
$$C = (2, 3, 5).$$
$$D = (1, 4, 6).$$

Find the volume of the parallelepiped determined by the vectors: AB, AC, and AD.

Solution:
To use the above formula for the volume, we need to compute the following vectors:

$$u = AB = (-1, 4, -3).$$
$$v = AC = (-3, 2, 2).$$
$$w = AD = (-4, 3, 3).$$

Then,

$$u.(v \times w) = \begin{bmatrix} -1 & 4 & -3 \\ -3 & 2 & 2 \\ -4 & 3 & 3 \end{bmatrix}$$

$$= 7(scalar).$$

Note: If u, and v are coplanar vectors, then $u.(v \times w) = 0$.

Projection of a Vector:

Vector u projected on vector v is denoted by , $Proj_v u$, given in formula as: (Where $0 < \theta < \pi/2$)

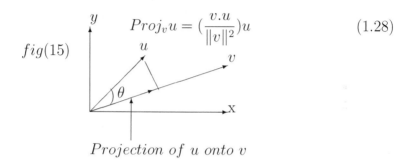

$$Proj_v u = (\frac{v.u}{\|v\|^2})u \tag{1.28}$$

$fig(15)$

Projection of u onto v

To derive equation (1.28), we start with,

$$Proj_v u = comp_v \frac{v}{\|v\|}$$

$$Comp_v u = \frac{u.v}{\|v\|}$$

then,

$$Proj_u = \frac{u.v}{\|v\|}.\frac{v}{\|v\|} = \frac{u.v}{\|v\|^2}.v$$

Then the normal projection will be :

$$Proj_N u = u - Proj_v u = u - (\frac{u.v}{\|v\|^2}).v. \tag{1.29}$$

This normal component can also be thought of as follows:

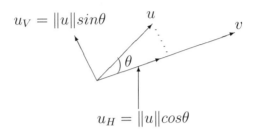

$$u_V = \|u\| sin\theta$$

$$u_H = \|u\| cos\theta$$

Vector u can be resolved into 2-components:

1. Horizontal components (u_H) which is the projection of u onto v , or the $x - axis$, found as $Proj_v u = (\frac{u.v}{\|v\|^2})v$, and,

2. Vertical component (u_V) which is the projection of u onto y-axis. Then the vector u in terms of components can be written as:

$$
\begin{aligned}
u &= u_H + u_v. \\
Or, u_V &= u - u_H . Then, \\
u_V &= u - (\frac{u.v}{\|v\|^2})v.
\end{aligned}
$$

Projection of vectors are used in physics to find the amount of work done along a distance, where: work$(W) = F.d$, d is the displacement along the direction of motion.

Note: If $\pi/2 < \theta < \pi$, then the projection will be as shown:

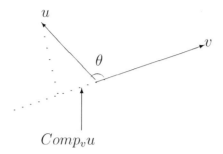

$$Comp_v u = \|u\| cos\theta.$$
$$But, \ cos\theta = \frac{u.v}{\|u\|\|v\|}.$$
$$Then, \ Comp_v u = \|u\|\frac{u.v}{\|u\|\|v\|} = \frac{u.v}{\|v\|}.$$

But u is projected on vector v, and its projection can be formed by multiplying the component of u along v by a unit vector in the direction of v:

$$Proj_v u = (Comp_v u) \times (unit \, vector \, of \, v).$$

$$= \frac{u.v}{\|v\|} \cdot \frac{v}{\|v\|} = (\frac{u.v}{\|v\|^2})^2$$

Then,

$$Proj_v u = (\frac{u.v}{\|v\|^2})v.$$

Corollary:

The vectors u, and v in $\mathbf{R^3}$ are parallel, if and only if, $u \times v = 0$. taking the nontrivial case, where both u, and v are nonzero vectors, then u, and v are parallel if:

1. $\theta = 0 \longrightarrow sin0 = 0 \longrightarrow u \times v = 0$
2. $\theta = \pi \longrightarrow sin\pi = 0 \longrightarrow u \times v = 0.$

1.7   Exercise - 3

For problems (1-3) find $u \times v$:
1. $u = <4, 2, 1>$, $v = < 2, 4, 3>$.
2. $u = 5i$, $v = 4i + 5j$.
3. $u = <4, 1, 4>$, $v = <-1, 6, 1>$.

In problems (4-6), find a vector orthogonal to both vectors:
4. $u = (2, 3, -1)$, $v = (4, 6, 1)$.
5. $u = 5i - 3k$, $v = 3i + 6j + 5k$.
6. $u = (4i-5j+k)$, $v = (5i +k)$.

In problems (7-9) for the given vectors:
$u = 4i + j + 3k$, $v = 3i + 4j - k$, and $w = 5i + 4j + k$: Compute
the value of the given expressions:
7. $u \cdot (v \times w)$.
8. $u \times (v \times w)$.
9. $(u \times v) \cdot (u \times w)$

10.   Find the area of the parallelogram with adjacent sides
AB, AC, with the points $A, B,$ and C given as: $A = (5, 3, 2), B = (4, 4, 1),$ and $C = (6, 2, 4)$

11. Find the area of the triangle with the following vertices:

$$A = (6, 0, 5).$$
$$B = (8, 3, 1).$$
$$C = (7, 4, 5).$$

12.   Find the volume of the parallelepiped with the adjacent
sides AB, AC, and AD and the following given points:

$$A \;=\; (5, 4, -3).$$

$$B = (3, 6, 0).$$
$$C = (0, 5, 2).$$
$$D = (6, 5, -6).$$

13. Show that the following three vectors are co planner:

$$u = 4i - 6j + 8k.$$
$$v = 2i + 4j - 2k, and$$
$$w = 14i + 10k, \ are \ coplanner.$$

14. Show that the given points lie in the same plane.

$$A = (3, 1, 4).$$
$$b = (5, -2, 3).$$
$$c = (2, 3, 4).$$
$$d = (3, 2, 3).$$

15. Find a vector orthogonal to the given vectors:

$$u = 2j - 6k,$$
$$v = 4i + 8j - 2k.$$

16. Find the angle between the two vectors from problem (15).

17. Find the projection of vector $u = -3i - j - k$ onto vector $v = 4i + 2j - k$. See solutions on page-168.

1.8   Lines and Planes

In the previous sections we have described the vector as a directed line segment with direction . This means vectors and lines are related to each other. In this section we will use vectors to describe lines. First, we will start with equations of lines, the basic equations used in Algebra are:

1. The slope-intercept equations,

$$y = mx + b, \qquad (1.30)$$

where, m = slope, and b = y-intercept.

2. The general equation:

$$ax + by + cz = 0, \qquad (1.31)$$

where, a, b, and c are constants.

A line can also be represented by parametric equation that relates both x, and y to t.

Equations of a line in a space :

1. Vector Equations:

A line in the space can be described by a point on the line, and a direction for the line, that is parallel to the line. Suppose L is a line that passes through $p_0(x_0, y_0, z_0)$ and has a direction vector $u = (u_1, u_2, u_3)$, or u has initial point p_0.

Then a point $p(x, y, z)$ will be on the line L, if and only if, $\overline{p_0p} = tu$.

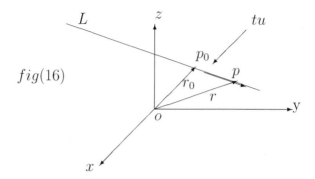

$fig(16)$

$\overline{p_0p} = r - r_0$, $-\infty < t < \infty$. But $\overline{p_0p} = tu$.

Thus the **vector equation** for the line L is:

$$\overline{p_0p} = r - r_0 = tu. \tag{1.32}$$

Where, $r = \overline{op}$, and $r_0 = \overline{op_0}$.

Example-1.11:
Find the vector equation of the line L passing through the point $(2, 7, -2)$, and parallel to the vector $(5, 1, 3)$.

Solution:
The line L is passing through the point $(2, 7, -2)$ which means, $r_0 = (2, 7, -2)$, and vector $u = (5, 1, 3)$.
then the vector equation for the line L is:

$$
\begin{aligned}
r &= (2, 7, -2) + t(5, 1, 3) \\
&= (2 + 5t, 7 + t, -2 + 3t).
\end{aligned}
$$

Example- 1.12:

A line L is passing through two points:

$$p_1 = (4, -1, 2).$$
$$p_2 = (-1, 2, 3).$$

Find the vector equation of the line L.

Solution:

Consider $p_1 = r_0 = (4, -1, 2)$, and the direction vector,
$u = \overline{p_1 p_2} = (-1 - 4, 2 + 1, 3 - 2) = (-5, 3, 1)$.
Thus the vector equation for L is:

$$
\begin{aligned}
r &= (4, -1, 2) + t(-5, 3, 1) \\
  &= (4 - 5t, -1 + 3t, 2 + t)
\end{aligned}
$$

2. Parametric equations of the line:

the vector equations derived above can be written in terms of x, y, and z, as follows:

$$
\begin{aligned}
(x, y, z) &= (x_0, y_0, z_0) + t(u_1, u_2, u_3) \\
          &= (x_0 + tu_1, y_0 + tu_2, z_0 + tu_3).
\end{aligned}
$$

This gives the following components:

$$x = x_0 + u_1 t,$$
$$y = y_0 + u_2 t,$$
$$z = z_0 + u_3 t,$$

For $-\infty < t < \infty$. These are called the parametric equations for the line L.

Example- 1.13:

From the given parametric equations of the line:

$$x = -5t + 1.$$
$$y = 4t + 2.$$

Graph and find :

a) Slope-intercept equation for the line.

b) The general equation of the line.

Solution:

To graph the line we need a point and a slope. Let:

$$Let\ t = 0 \Longrightarrow x = 1, y = 2 \Longrightarrow p_0 = (1, 2).$$
$$Let\ t = 1 \Longrightarrow x = -4, y = 6 \Longrightarrow p_0 = (-4, 6).$$

The graph of the line is as shown:

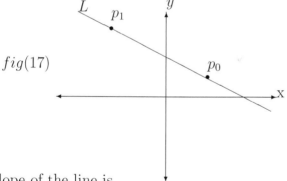

$fig(17)$

The slope of the line is,

$$m = \frac{y_2 - y_1}{x_2 - x_1} = \frac{6 - 2}{-4 - 1} = \frac{4}{-5}. \tag{1.33}$$

b is the y-intercept, where $x = 0$. Thus letting $x = 0$, and solving for t gives $t = 1/5$, then substituting to get y gives $y = 14/5 = b$. Then,

a) The slope-intercept equation of the line is:
$y = -4/5x + 14/5$.
b) The general equation of the line is: $4x + 5y - 14 = 0$.

Example- 1.14:
A line L passing through two points $p_1 = (2, -1, 2)$, and
$p_2 = (5, 2, -2)$. Find the parametric equation for the line L.

Solution:
As in the previous example, we can find the line by using the
equation :
$r = r_0 + ut$, where, $r_0 = (5, 2, -2)$, and the direction vector,
$u = \overline{p_1 p_2} = p_2 - p_1 = (3, 3, -4)$. Then equation of the line
becomes:

$$\begin{aligned} r &= (5, 2, -2) + t(3, 3, -4) \\ &= (5 + 3t, 2 + 3t, -2 - 4t) \end{aligned}$$

This gives the following parametric equations(components):

$$x = 5 + 3t,$$
$$y = 2 + 3t,$$
$$z = -2 + 5t,$$

For $-\infty < t < \infty$.

There are several parametric representation for the line, the
following example illustrates that.

Example- 1.15:
Find the parametric representation through the points:
$p_1 = (x_1, y_1) = (1, 4)$.
$p_2 = (x_2, y_2) = (-4, 8)$.

Solution:

1. The elementary representation: with Slope-Intercept

$$
\begin{aligned}
Slope = m \ &= \ \frac{y_2 - y_1}{x_2 - x_1} \\
&= \ \frac{8 - 4}{-4 - 1} = \frac{3}{-5}.
\end{aligned}
$$

Equation of the line is;

$$
\begin{aligned}
y - y_1 \ &= \ m(x - x_1). \\
Then, \ y - 4 \ &= \ -\frac{3}{5}(x - 1). \\
Or, \ y \ &= \ -\frac{3}{5}x + \frac{23}{5}.
\end{aligned}
$$

2. Setting $x = t$ gives the parametric equation for the line as:

$$
\begin{aligned}
x \ &= \ t. \\
y \ &= \ -\frac{3}{5}t + \frac{23}{5}.
\end{aligned}
$$

3. Using parametric equation as:

$$
\begin{aligned}
x \ &= \ x_1 + t(x_2 - x_1) = 1 - 5t. \\
y \ &= \ y_1 + t(y_2 - y_1) = 4 + 4t.
\end{aligned}
$$

If the points are switched, we still get the same line:

$$
\begin{aligned}
x \ &= \ x_2 + t(x_1 - x_2) = -4 + 5t. \\
y \ &= \ y_2 + t(y_1 - y_2) = 8 - 4t.
\end{aligned}
$$

3. Symmetric equations of the line:

Let the direction angle of the line L be α, β, γ, and let $p_0(x_0, y_0, z_0)$ be a point of the line.

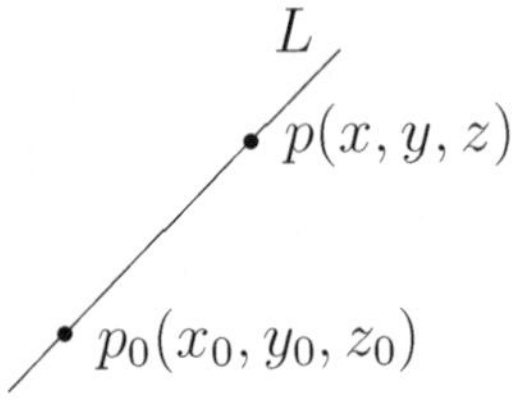

Then L is the locus of $p(x, y, z)$ moving so that:

$$
\begin{aligned}
x - x_0 &= t\cos\alpha. \\
y - y_0 &= t\cos\beta. \\
z - z_0 &= t\cos\gamma. \\
Or, \quad x &= x_0 + t\cos\alpha. \\
y &= y_0 + t\cos\beta. \\
z &= z_0 + t\cos\gamma.
\end{aligned}
$$

Where, the parameter t represents the length $p_0 p$. If a, b, c are direction numbers of L, then we get the following parametric equations:

$$
\begin{aligned}
x &= x_0 + at, \\
y &= y_0 + bt, \\
z &= z_0 + ct,
\end{aligned}
$$

Solving the above equations for t and equating them gives the symmetric equation for the line:

$$
\frac{x - x_0}{a} = \frac{y - y_0}{b} = \frac{z - z_0}{c} \tag{1.34}
$$

Where, a, b, and c are direction numbers of the line.

Example- 1.16:
a) Find the symmetric equation for the line L with direction vector : $(-4, 8, 6)$, and contains the point: $(6, -2, 8)$.
b) Find the parametric equations for the line L in (1).

Solution:
a) The symmetric equation is :

$$\frac{x-6}{-4} = \frac{y+2}{8} = \frac{z-8}{6} \qquad (1.35)$$

b) From the equation : $r = (6 - 4t, -2 + 8t, 8 + 6t)$ we get the following parametric equations:

$$x = 6 - 4t$$
$$y = -2 + 8t.$$
$$z = 8 + 6t.$$

Example- 1.17:
Find the symmetric equation of the line L with the point $p_0 = (8, -2, 2)$, and parallel to the vector $u = 4i - 6j$.

Solution:
The parametric equations for the line are:

$$x = x_0 + u_1 t = 8 + 4t.$$
$$y = y_0 + u_2 t = -2 - 6t.$$
$$z = z_0 + u_3 t = 2,$$

Solving the above parametric equations for t, then letting $t = 1$, we get the symmetric following equation:

$$\frac{x-8}{4} = \frac{y+2}{-6}, z = 2. \qquad (1.36)$$

The angle between two lines in R^3:

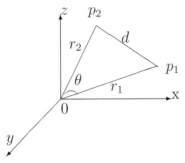

For the given two lines in the figure, let:

$$r_1 = op_1.$$
$$r_2 = op_2.$$
$$d = p_1 p_2.$$

Also, consider the followings:
$\alpha_1 =$ the angle between r_1 and x_1.
$\beta_1 =$ the angle between r_1 and y_1.
$\gamma_1 =$ the angle between r_1 and z_1.

Where, $r_1^2 = x_1^2 + y_1^2 + z_1^2$.

Similarly for r_2:
$\alpha_2 =$ the angle between r_2 and x_2.
$\beta_2 =$ the angle between r_2 and y_2.
$\gamma_2 =$ the angle between r_2 and z_2.

Where, $r_2^2 = x_2^2 + y_2^2 + z_2^2$.

$$Then, \ d^2 = (x_2 - x_1)^2 + (y_2 - y_1)^2 + (z_2 - z_1)^2.$$

Using Cosine-Law gives:

$$cos\,\theta = \frac{r_1^2 + r_2^2 - d^2}{2r_1 r_2}.$$

$$Or, \ cos\theta = \frac{x_1^2 + y_1^2 + z_1^2 + x_2^2 + y_2^2 + z_2^2 - d^2}{2r_1 r_2}.$$

$$Simplifying \ gives, \ cos\theta = \frac{x_1 x_2 + y_1 y_2 + z_1 z_2}{r_1 r_2}.$$

Then, $cos\theta = cos\alpha_1 cos\alpha_2 + cos\beta_1 cos\beta_2 + cos\gamma_1 cos\gamma_2.$

Where, $cos\alpha_1 = \frac{x_1}{r_1}, cos\alpha_2 = \frac{x_2}{r_2}$...etc.

1. If the two lines are parallel $\Rightarrow \theta = 0 \Rightarrow cos\theta = 1$, and $\alpha_1 = \alpha_2$, $\beta_1 = \beta_2$, and $\gamma_1 = \gamma_2$, then

$$Cos^2\alpha + cos^2\beta + cos^2\gamma = 1.$$

2. If the two lines are perpendicular, then $\theta = 90° \Rightarrow cos\theta = 0$, and ,

$$cos\alpha_1 cos\alpha_2 + cos\beta_1 cos\beta_2 + cos\gamma_1 cos\gamma_2 = 0.$$

Example- 1.18:
Find the distance d between the two points $p_1 = (6, -1, 4)$, and $p_2 = (-3, 5, 7)$.
Solution:

$$
\begin{aligned}
d &= \sqrt{(x_2 - x_1)^2 + (y_2 - y_1)^2 + (z_2 - z_1)^2}. \\
&= \sqrt{(-3 - 6)^2 + (5 + 1)^2 + (7 - 4)^2}. \\
Then, \ d &= \sqrt{126}.
\end{aligned}
$$

Example- 1.19:
Find the direction cosines and the direction angles of the line drawn from the origin to the point $(-3, 5, 7)$.
Solution:

$$cos\,\alpha = \frac{x_2 - x_1}{\sqrt{(x_2 - x_1)^2 + (y_2 - y_1)^2 + (z_2 - z_1)^2}} = \frac{-3}{\sqrt{83}}.$$

Then, $\alpha \approx 109.23°.$

$$\cos \beta = \frac{y_2 - y_1}{\sqrt{(x_2 - x_1)^2 + (y_2 - y_1)^2 + (z_2 - z_1)^2}} = \frac{5}{\sqrt{83}}.$$

Then, $\beta \approx 56.7°$.

$$\cos \gamma = \frac{z_2 - z_1}{\sqrt{(x_2 - x_1)^2 + (y_2 - y_1)^2 + (z_2 - z_1)^2}} = \frac{7}{\sqrt{83}}.$$

Then, $\gamma \approx 39.8°$.

1.9   Exercise - 4

A line L ,containing the given points, is parallel to the vector u (given). Find the Vector equation and symmetric equation:

1. $P = (-3, 2, 1);$ $u = -3i + j + 2k$.

2. $p = (3, 0, 6);$ $u = 3j + 4k$.

3. $P = (-3, 2, 1);$ $u = -3i + j$

4. Find the parametric equation for the line with the point $p = (3, -2, 2)$ and is parallel to the line with equation:

$$\frac{x - 2}{5} = \frac{y + 4}{5} = z.$$

5. Two lines L_1 with two points:

$$
\begin{aligned}
P_1 &= (4, 3, 4), \\
P_2 &= (-3, -5, 7),
\end{aligned}
$$

and L_2 with two points:

$$P_1 = (-1, 3, 3),$$
$$P_2 = (-4, 5, 2),$$

Show that the two lines are perpendicular.

6. A line with the symmetric equation :

$$\frac{x-1}{-3} = \frac{y-2}{12}, \ and \ z = 5.$$

And a second line with two points: $p_1 = (5, 7, 9)$, and $p_2 = (4, 11, 9)$.

Show that the two lines are skewed. See solutions on page-171.

1.10 Equations of the Plane in the Space

As seen in the previous sections the x, y, and z components of a line were expressed in terms of a single parameter t, so the line in the space is a one-dimensional. But the plane in the space is a two-dimensional, so its components must be expressed in two-parameters say: s, and t. The line was determined by : initial position vector (r_0), and direction vector (u). Now the plane is determined by its initial position vector, and two-non parallel direction vectors v, and w that lie in the plane. Thus the vector parameter form of the plane is:

$$u + sv + tw \qquad (1.37)$$

A General equation for the Plane

Let Ω be a plane in R^3, containing the 2-points : $p_0 = (x_0, y_0, z_0)$, and $p = (x, y, z)$; the direction vector $\overline{p_0 p}$, and normal vector $n = ai + bj + zk$. then using the dot-product rule :

$$n.\overline{p_0 p} = 0. \qquad (1.38)$$

But, $\overline{p_0 p} = r - r_0$, then equation (1.39) can be written as:

$$n.(r - r_0) \;=\; (a, b, c).(x - x_0, y - y_0, z - z_0) = 0.$$

Then from this we get the general equation of the plane as :

$$a(x - x_0) + b(y - y_0) + c(z - z_0) = 0 \qquad (1.39)$$

For a given initial point $p_0 = (x_0, y_0, z_0)$ the above equation becomes:

$$ax + by + cz + d = 0. \qquad (1.40)$$

In the plane Ω suppose we choose two pints such as:

$$p_1 \;=\; (x_1, y_1, z_1)$$
$$P_2 \;=\; (x_2, y_2, z_2)$$

The direction vector : $\overline{P_1 P_2} = ((x_2 - x_1), (y_2 - y_1), (z_2 - z_1)$
From the equation of the plane (1.40) we have:

$$ax_1 + by_1 + cz_1 + d \;=\; 0$$
$$ax_2 + by_2 + cz_2 + d \;=\; 0$$

Subtracting the first equation from the second gives:
$a(x_2 - x_1) + b(y_2 - y_1) + c(z_2 - z_1 0 = 0$
This last equation is the dot-product of the vector (a,b,c) and direction vector $\overline{p_1 p_2}$.

Example-1.20:
Find the equation of the plane passing through the points:
$P = (-1, 0, 3), Q = -2, 5, 6), R = (-6, -3, 4)$.
Solution:

$$Let, \ u \ = \ PQ = Q - P = < -1, 5, 3 > .$$
$$And \ v \ = \ PR = R - P = < -5, -3, 1 > .$$

$$u \times v \ = \ \begin{bmatrix} i & j & k \\ -1 & 5 & 3 \\ -5 & -3 & 1 \end{bmatrix}$$
$$= \ 14i - 14j + 28k.$$

Let the normal vector be $n = 14i - 14j + 28k = 14 < 1, -1, 2 >$.
Choosing anyone of the 3-given points: Say$(-1, 0, 3) = (x_0, y_0, z_0)$.
Then using the general plane equation:
$a(x - x_0) + b(y - y_0) + c(z - z_0) = 0$ gives the equation of the plane:

$$(x + 1) - (y - 0) + 2(z - 3) \ = \ 0.$$
$$Or, \ x - y + 2z = 5.$$

Where, $(a, b, c) = < 1, -1, 2 >$.

Example-1.21:
A plane containing initial point $p_0 = (2, 4, 5)$, and a direction vectors:
$v = j + k, u = i - j$. Find the vector parameter form of the plane.
Solution:
From the given information, $p_0 = (2, 4, 5); v = (0, 1, 1);$
$w = (1, -1, 0)$, we can write the parametric components as:

$$x(s, t) \ = \ 2 + t.$$

$$y(s,t) = 4 + s - t.$$
$$z(s,t) = 5 + s.$$

Thus the vector parameter form for the plane is:
$$u + sv + tw = (2, 4, 5) + s(0, 1, 1) + t(1, -1, 0)$$

Equation of the Normal distance - D:

Let θ be the angle between the vector u, and the direction vector $\overline{p_0 p_1}$ such that $0 < \theta < \pi$, then,

$$D = \|\overline{p_0 p_1}\| sin\theta \qquad (1.41)$$

But,
$$\|u \times \overline{p_0 p_1}\| = \|u\| \|\overline{p_0 p_1}\| sin\theta. \qquad (1.42)$$

Then, we get the equation of the distance as:

$$D = \frac{\|u \times \overline{p_0 p_1}\|}{\|u\|} \qquad (1.43)$$

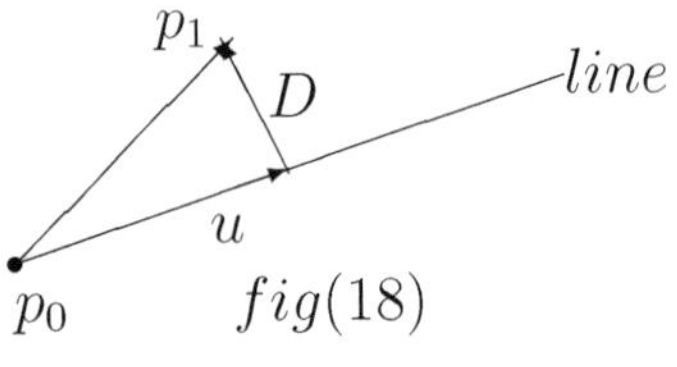

$fig(18)$

Equation(1.43) is the equation of the normal distance.

Example- 1.22:
Find the normal distance D from the point : $p_1 = (2, 1, -1)$ to the line with the parametric equations given as :
$r = (3t, 1 + 2t, -5 - t)$.

Solution:
From the parametric equations we get the following components:

$$
\begin{aligned}
x &= 3t, \\
y &= 1 + 2t, \\
z &= -5 - t.
\end{aligned}
$$

From these components we get the directional equation for the line $u = (3, 2, -1)$, and $p_0 = (0, 1, -5)$ as a point on the line. Then $\overline{p_0 p_1} = (2, 0, 4)$. Then using the distance formula we can compute D as follows:

$$
\begin{aligned}
D &= \frac{\|u \times \overline{p_0 p_1}\|}{\|u\|} \\[2mm]
&= \left\| \frac{(3, 2, -1) \times (2, 0, 4)}{(3, 2, -1)} \right\|
\end{aligned}
$$

$$
Then, \quad D = \left\| \frac{\begin{pmatrix} i & j & k \\ 3 & 2 & -1 \\ 3 & 0 & 4 \end{pmatrix}}{\sqrt{14}} \right\|.
$$

$$
Or, \quad D = \left\| \frac{8i - 15j - 6k}{\sqrt{14}} \right\| = \sqrt{\frac{325}{14}}.
$$

Example- 1.23:
Find the equation of the plane with normal vector $u = 6i + 7k$, and contains the point $p_0 = (-1, 3, 6)$.

Solution:
From the given point we have: $x_0 = -1, y_0 = 3, z_0 = 3$, and from the normal vector we have : $u = 6, v = 0, w = 7$. Substituting in the General equation of the plane we get:

$$u(x - x_0) + v(y - y_0) + w(z - z_0) = 0.$$
$$6(x + 1) + 0(y - 3) + 7(z - 3) = 0.$$
$$6x + 7z - 15 = 0.$$

Example- 1.24:
Find the equation of the vector that is normal to the plane : $2x - 2y + 6z = -30$.

Solution:
Comparing the given equation with the plane with general equation we get the normal vector as follows:

$$ax + by + cz = d$$
$$2x - 2y + 6z = -30.$$

Then, the normal vector $N = 2i - 2j + 6k$.

Example- 1.25:
A plane containing three points:

$$p_1 = (2, 0, 3)$$
$$p_2 = (-2, 3, 4)$$
$$p_3 = (5, 2, 7).$$

Find the equation of the plane.

Solution: Let,

$$u = \overline{p_1 p_2} = (-2 - 2, 3 - 0, 4 - 3) = (-4, 3, 1), and,$$
$$v = \overline{p_1 p_3} = (5 - 2, 2 - 0, 7 - 3) = (3, 2, 4).$$

Then,

$$N = \overline{p_1 p_2} \times \overline{p_1 p_3}$$

$$= \begin{pmatrix} i & j & k \\ -4 & 3 & 1 \\ 3 & 2 & 4 \end{pmatrix}$$

$$= 10i + 19j - 17k$$

Taking a point say $p_1 = (2, 0, 3)$, this point lie on the plane, and since N is normal to the plane, then it is normal to any point on the plane including p_1. Then the equation of the plane can be written as:

$$10(x - 2) + 19y - 17(z - 3) = 0.$$
$$10x + 19y - 17z = -31$$

Note: In terms of the normal vector N the equation of the distance D can be written as:

$$D = \frac{\|N . \overline{p_0 p_1}\|}{\|N\|} \tag{1.44}$$

> **Proof of (1.44) is left for the students.**

Example-1.26:
Compute the distance D between the point $p_1 = (-2, 2, 4)$, and the plane with equation $6x - 4y + 2z = 2$.

Solution:
From the given equation of the plane , we get the equation of the normal vector $N = 6i - 4j + 2k$. Choosing another point on

the plane by setting : $x = y = 0$, and $z = 1$, let $p_0 = (0, 0, 1)$, the direction vector is: $\overline{p_0 p_1} = (-2, 2, 3)$. Using equation (1.44) we get the distance:

$$D = \frac{\|N.\overline{p_0 p_1}\|}{\|N\|} = \frac{4}{\sqrt{56}}$$

1.11   Curvilinear Coordinates

Types of Coordinate Systems:

1. Cartesian Coordinate System:

A point in this system is located in R^2 by a pair of numbers (x, y), and in R^3 by a triple of numbers (x, y, z). The axes are maintained in a right-handed configuration (i.e. thumb up represents the z-axis, and the 4-fingers curled in represents x, and y axis).

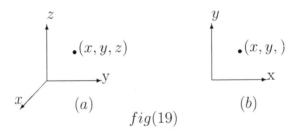

$fig(19)$

2. Polar Coordinate System:

In this system a fixed point is taken as a reference called the pole o, and a ray emanating from the pole, called the polar axis directed horizontally to the right. Taking any point in the plane say p, then $op = r$ is the length , and direction is θ.

$fig(20)$ $p(r,\theta)$ θ x

In terms of Polar the Cartesian coordinates are:

$$x = r\cos\theta.$$
$$y = r\sin\theta$$

Changing to polar coordinates:

$$\begin{aligned} x^2 + y^2 &= r^2\cos^2\theta + r^2\sin^2\theta \\ &= r^2(\cos^2\theta + \sin^2\theta) \\ &= r^2 \\ \tan\theta &= \frac{y}{x}, \ where \ x \neq 0. \end{aligned}$$

3. Cylindrical Coordinate System: In this system the coordinates are Length of r, direction θ, and $z-axis$. In terms of Cylindrical, the Cartesian coordinates are:

$$x = r\cos\theta.$$
$$y = r\sin\theta$$
$$z = z.$$

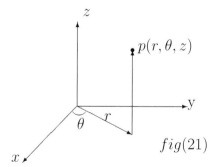

$fig(21)$

4. Spherical Coordinate System: The point in this system is (ρ, θ, ϕ):

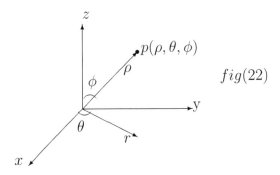

$fig(22)$

The cylindrical coordinates are:

$$
\begin{aligned}
r &= \rho \sin\phi. \\
\theta &= \theta \\
z &= \rho \cos\phi, \\
\rho^2 &= r^2 + z^2 \\
\tan\phi &= \frac{r}{z} \\
\theta &= \theta.
\end{aligned}
$$

Their Cartesian coordinates are:

$$
\begin{aligned}
x &= \rho \sin\phi \cos\theta. \\
y &= \rho \sin\phi \sin\theta, \\
z &= \rho \cos\phi, \\
\rho^2 &= x^2 + y^2 + z^2. \\
\tan\theta &= \frac{y}{x}. \\
\tan\phi &= \frac{\sqrt{x^2 + y^2}}{z}.
\end{aligned}
$$

Example-1.27:
A point given in polar coordinates: $p = (3, \pi/3)$. Give the point in rectangular coordinates.

Solution:
The given point in polar coordinates as: $p(r, \theta) = (3, \pi/3)$, means $r = 3$, and $\theta = \pi/3$. Then in rectangular coordinates this will be changed as:

$$x = r\cos\theta = 3\cos\pi/3 = 3\cos 60 = 3/2.$$

$$y = r\sin\theta = 3\sin\pi/3 = 3\left(\frac{\sqrt{3}}{2}\right).$$

Example-1.28:
A point with rectangular coordinates $(\sqrt{3}, 2)$. Transform the point into polar coordinates.

Solution:
In rectangular coordinates: $(x, y) = (\sqrt{3}, 2)$, then the polar coordinates:
$p(r, \theta)$ are, $r = x^2 + y^2 = 7$, and $\theta = than^{-1}(\frac{y}{x}) = 30^0$.

Example- 1.29:
Transform the equation $, r = 2\cos\theta$ into rectangular coordinates.

Solution:
Multiplying the given equation by r gives, $r^2 = 2r\cos\theta = 2x$. Then $x^2 + y^2 = r^2 = 2x$.

Example-1.30:
Convert The rectangular point $(x, y, z) = (-1, \sqrt{3}, 2)$, to Spherical coordinates (ρ, θ, z).

Solution:

$$\rho = \sqrt{x^2 + y^2 + z^2} = 2\sqrt{2}.$$

$$\phi = \cos^{-1}(\frac{z}{\rho}) = \cos^{-1}(\frac{2}{2\sqrt{2}}) = \frac{\pi}{4} = 45°.$$

$$And, \ \cos\theta = \frac{x}{\rho \sin \phi} = -\frac{1}{2}.$$

$$Then, \ \theta = \cos^{-1}(-\frac{1}{2}) = 120° = \frac{2\pi}{3}.$$

Then $(-1, \sqrt{3}, 2) \Rightarrow (2\sqrt{2}, \frac{2\pi}{3}, \frac{\pi}{4})$.

Example-1.31:
Transform the equation , $x^2 - y^2 - 6y = 0$, into polar coordinate.

Solution:

$$x^2 - y^2 - 6y = 0.$$
$$x^2 - y^2 = 6y.$$
$$r^2 \cos^2\theta - r^2 \sin^2\theta = 6r\sin\theta.$$
$$r^2(\cos^2\theta - \sin^2\theta) = 6r\sin\theta.$$

But $\cos^2\theta - \sin^2\theta = \cos2\theta$. Then the transformation is:

$$r^2 = 6r \ \sin\theta.$$
$$r = \frac{6 \sin\theta}{\cos(2\theta)}.$$

Example-1.32:
Change the spherical point: $(2, \pi/2, \pi/6)$ into a rectangular point.

Solution:
The spherical point $= (\rho, \phi, \theta) = (2, \pi/2, \pi/6)$.

Where, $\rho = 2$, $\phi = \pi/2 = 90^0$, and $\theta = \pi/6 = 30^0$. Then the Cartesian coordinates are:

$$x = 2sin90 \ cos30 = \sqrt{3}.$$
$$y = 2sin90 \ sin30 = 1.$$
$$z = 2cos90 = 0.$$

Then the point in rectangular coordinates is : $(\sqrt{3}, 1, 0)$

Example -1.33:
Change the given Cartesian coordinate point: $(1, \sqrt{3}, 7)$ into cylindrical point.

Solution:
From the given point we get $r = 1$, $\theta = \sqrt{3}$, and $z = 7$. Then in cylindrical coordinates it is: $r = \sqrt{x^2 + y^2} = 2$, $\theta = tan^{-1}(\frac{1}{\sqrt{3}}) = 30^0$, and $z = 7$.
Then the transformed cylindrical point is: $(r, \theta, z) = (2, 30^0, 7)$.

Example -1.34:
Convert the rectangular coordinates $(-1, \sqrt{3}, 2)$ to spherical coordinates.
Solution:
From the given Cartesian coordinates: $(x, y, z) = (-1, \sqrt{3}, 2)$ we have:
$x = -1, y = \sqrt{3}, z = 2$, we can find: $\rho = \sqrt{x^2 + y^2 + z^2} = 2\sqrt{2}$.
$\phi = cos^{-1}(z/\rho) = cos^{-1}(1/\sqrt{2}) = 45°$.

$cos\theta = \frac{x}{\rho sin\phi} = -1/2$, then $\theta = cos^{-1}(-1/2) = 120° = \frac{2\pi}{3}$.

Then the Cartesian coordinates $= (-1, \sqrt{3}, 2) \Rightarrow (2\sqrt{2}, \frac{2\pi}{3}, \frac{\pi}{4})$ = Spherical coordinates.

Curvilinear Coordinates:

1. Changing Cylindrical coordinates (r, θ, z) to rectangular Coordinates (x, y, z).

$$
\begin{aligned}
x &= r\cos\theta, \\
y &= r\sin|\,\theta, \\
z &= z. \\
r &= \sqrt{x^2 + y^2}, \\
And, \quad \tan\theta &= \frac{y}{x}.
\end{aligned}
$$

2. Changing spherical coordinates (ρ, θ, ϕ) to rectangular co-ordinates
(x, y, z).

$$
\begin{aligned}
x &= \rho\sin\phi\cos\theta. \\
y &= \rho\sin\phi\sin\theta. \\
z &= \rho\cos\phi. \\
\rho &= \sqrt{x^2 + y^2 + z^2}. \\
\cos\phi &= \frac{z}{\rho}, and \\
\phi &= \cos^{-1}(\frac{z}{\rho}, \ 0 < \phi < \pi. \\
\cos\theta &= \frac{x}{\rho\sin\theta}. \\
\sin\theta &= \frac{y}{\rho\sin\phi}.
\end{aligned}
$$

1.12   Exercise - 5

1. Find the Cartesian coordinates for the given point whose polar coordinates are $(\sqrt{2}, \pi/3)$.

2. Find the polar coordinates for the point whose Cartesian coordinates are $(3, 2)$.

3. A point in cylindrical coordinates is $(1/2, \pi/3, 2)$, transform the point into Cartesian coordinate.

4. Find the rectangular(Cartesian) coordinates of the point whose spherical coordinates are $(4, \pi/2, \pi/4)$.

5. Find the cylindrical coordinates for the given Cartesian coordinate point $(2, \sqrt{3}, 3)$.
See solutions on page-173.

1.13   Vector Valued Functions

Until now we have studied functions whose domain and range both consist of real numbers. Now we will deal with a new type of functions that have a domain of real numbers but a range of vector type, and these are called vector-valued-functions. Vector valued functions are used in solving many problems in space, such as describing the motion of a Satalite.

Definition:A vector valued function consists of two parts: a domain which is a collection of real numbers, and, a rule(range) which assigns values for each number.

We will be using t as a number in the domain, and the vector valued function will be represented by a bold-face letter such as: $\mathbf{r(t)}$, $\mathbf{F(t)}$, $\mathbf{x(t)}$, ... etc. As we have done before we will use 3-real numbers x, y, and z, and basis i, j, and k. So the vector function can be written as:

$$\mathbf{F(t)} = xi + yj + zk. \tag{1.45}$$

The components are each a function of t, where $a \leq t \leq b$:

$$
\begin{aligned}
x &= f_1(t), \\
y &= f_2(t), \\
z &= f_3(t).
\end{aligned}
$$

If $\mathbf{F}$ is continuous on the given interval, then the range of $\mathbf{F}$ is called a **curve**, t the **curve parameter**, and the interval the **parameter interval**.

Example-1.35:
Determine the domain and the component functions of the vector valued function: $\mathbf{F(t)} = t^2 i + t^3 j + t^4 k$.

Solution:
The function is defined everywhere in the interval $(-\infty, \infty)$, and the components of the function are:

$$
\begin{aligned}
f_1(t) &= t^2. \\
f_2(t) &= t^3. \\
f_3(t) &= t^4.
\end{aligned}
$$

Example- 1.36:
Determine the domain and the components of the given vector valued function: $\mathbf{f(t)} = (2ti - j) \times (2i - t^2 j + 2tk)$

Solution:
First we have to compute the cross-product to find the components of the vector valued function:

$$
\begin{aligned}
\mathbf{F(t)} &= \begin{bmatrix} i & j & k \\ 2t & -1 & 0 \\ 2 & -t^2 & 2t \end{bmatrix} \\
&= -2ti - 4t^3 j + (2 - 2t^3)k.
\end{aligned}
$$

we see from the last expression for the function , the domain of $\mathbf{F(t)}$ is $t \geq 0$, and the components of the function are:

$$
\begin{aligned}
f_1(t) &= -2t. \\
f_2(t) &= -4t^3. \\
f_3(t) &= 2 - 2t^3.
\end{aligned}
$$

Since it is difficult to graph the vector valued function, then its range is graphed usually. If we think of $\mathbf{F(t)}$ as a point in a space, then as t increases $\mathbf{F(t)}$ traces out a curve in a space. Thus the range of $\mathbf{F(t)}$ is associated with a curve in a space.

Example- 1.37:
For the vector valued function : $\mathbf{F}(\mathbf{t}) = (3t-2)i+(2t+1)j+4tk$.
Sketch the curve traced out by $\mathbf{F}$.

Solution:
In terms of parametric equations of a line, we write the components of the given function as:

$$x = 3t - 2 = x_0 + at.$$
$$y = 2t + 1 = y_0 + bt.$$
$$z = 4t = z_0 + ct.$$

Where, the point $(x_0, y_0, z_0) = (-2, 1, 0)$ is on the line that is parallel to the vector $(a, b, c) = (3, 2, 4)$

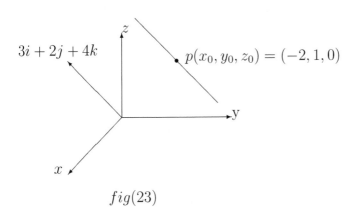

$3i + 2j + 4k$

$p(x_0, y_0, z_0) = (-2, 1, 0)$

$fig(23)$

1.14   Exercise - 6

determine the domain , and the components for each of the following vector functions:

1. $\mathbf{F(t)} = [(t-1)i + lnt\,j + sint\,k] \times [(2-t^2)i + e^5tj + t^{-2}k]$.

2. $\mathbf{F(t)} = \sqrt{t-5}i + \sqrt{t+4}j + 2k$.

3. Sketch the curve traced out by the vector valued function: $\mathbf{F(t)} = -3ti + (2t+1)j - (3t-2)k$.

4. Sketch the curve traced out by the vector valued function: $\mathbf{F(t)} = cost\,i + sint\,j,\,for\;0 \le t \le \pi/2$.

5. Sketch the curve traced out by the vector valued function: $\mathbf{F(t)} = cost\,i + sint\,j,\,for\;0 \le t \le 2\pi$.
See solutions on page-175.

1.15   Arc Length

In this section we will continue with our general study of parameterized curves in $\mathbf{R}^3$. Considering how to measure such
geometric properties as length and curvature. To measure the
length of a path: Let C be a space curve with finite length on
the interval $[a, b]$ for the vector function $f(t)$ with parameterized components:

$$x = f_1(t).$$
$$y = f_2(t).$$

And if $f_1'(t)$, and $f_2'(t)$ are continuous on the closed interval
$[a, b]$, then C is rectifiable and its length in 2-dimension is
given by:

$$L = \int_a^b \sqrt{(f_1(t))^2 + (f_2(t))^2} \; dt.$$
$$= \int_a^b \sqrt{(\frac{dx}{dt})^2 + (\frac{dy}{dt})^2} dt.$$
$$= \int_a^b \|f'(t)\| dt.$$

Theorem: Let C be the graph of the continuous vector function f on the interval $[a, b]$, and suppose f' is also continuous
on the interval $[a, b]$. Then C is recifiable , and its length L is
given by :

$$L = \int_a^b \|f'(t)\| dt. \tag{1.46}$$

Since C is a plane curve then its length is:

$$L = \int_a^b \sqrt{(f_1'(t))^2 + (f_2'(t))^2 + (f_3'(t))^2} \; dt.$$
$$= \int_a^b \sqrt{(\frac{dx}{dt})^2 + (\frac{dy}{dt})^2 + (\frac{dz}{dt})^2} \; dt.$$

Example- 1.38:

Find the length of the vector function : $\mathbf{F(t)} = (cost, sint, t)$ when t falls on the interval $[0, \pi]$.

Solution:

The components of f and their first derivatives with respect to t are as follows:

$$\begin{aligned} f_1 &= cost, \ f_1' = -sint. \\ f_2 &= sint, \ f_2' = cost. \\ f_3 &= t, \ f_3' = 1 \end{aligned}$$

Applying equation (1.46) gives:

$$\begin{aligned} L &= \int_0^\pi \|f'(t)\| dt. \\ &= \int_0^\pi \sqrt{(f_1')^2 + (f_2')^2 + (f_3')^2} \ dt. \\ &= \int_0^\pi \sqrt{(-sint)^2 + (cost)^2 + (1)^2} \ dt. \\ &= \int_0^\pi \sqrt{2} \ dt. \\ &= \sqrt{2}\pi. \end{aligned}$$

Example- 1.39:

Find the length of the curve on the closed interval $[-1, 3]$. $\mathbf{F(t)} =< (93 - 2t), (4 + 6t), (5 - 3t) >$.

Solution:

As done in the previous example, we take the first derivative of the components of f, then apply equation (1.45) to get:

$$\begin{aligned} L &= \int_{-1}^3 \sqrt{(f_1')^2 + (f_2')^2 + (f_3')^2} \ dt. \\ &= \int_{-1}^3 \sqrt{(-2)^2 + (6)^2 + (-3)^2} \ dt. \\ Then, \ L &= 28. \end{aligned}$$

Use of the Arc Length:

An important use of the Arc-Length is to find the Curvature, which is the measure of how sharply a curve bends . The curvature is usually labeled as **k**, also the unit tangent vector **T**, and the unit normal vector **N**. Here we will give their formula's without derivation.
The formula for the curvature is:

$$\mathbf{k} = \left\| \frac{d\mathbf{T}}{ds} \right\| = \frac{\|\mathbf{T}'(t)\|}{\|\mathbf{r}'(t)\|} = \frac{\|\mathbf{r}'(t) \times \mathbf{r}''(t)\|}{\|\mathbf{r}'(t)\|^3}. \tag{1.47}$$

The formula for the unit tangent vector is:

$$\mathbf{T} = \frac{\mathbf{r}'(t)}{\|\mathbf{r}'(t)\|}. \tag{1.48}$$

And the unit normal vector formula is (see fig. below):

$$\mathbf{N} = \frac{\mathbf{T}'(t)}{\|\mathbf{T}'(t)\|} = \frac{\frac{d\mathbf{T}}{dt}}{\left\|\frac{d\mathbf{T}}{T}\right\|}. \tag{1.49}$$

These formula's are used in Physics and in space.

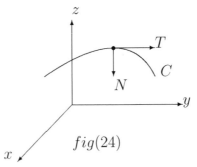

$fig(24)$

Fig(24) shows both tangent vector and normal vector on the curve C.

Example- 1.40:

Find **T**, **N**, and **k** at an arbitrary point on the curve with parametric components : $x = 2cos2t, y = 2sin2t, z = 4t$.

Solution:

We need to find $f(t), f'(t), f''(t)$, then apply the formula's to get the required quantity.

$$\begin{aligned}
f(t) &= 2cos2t, 2sin2t, 4t. \\
f'(t) &= -4sin2t, 4cos2t, 4. \\
f''(t) &= -8cos2t, -8sin2t, 0.
\end{aligned}$$

Then,

$$\|f'(t)\| = \sqrt{16sin^2(2t) + 16cos^2(2t) + 16} = 5.7$$

And,

$$T = \frac{f'(t)}{\|f'(t)\|} = (\frac{-4}{\sqrt{32}}sin2t, \frac{4}{\sqrt{32}}cos2t, \frac{16}{\sqrt{32}})$$

Then to find N, we need to compute $T'(t)$ first:

$$T'(t) = (\frac{-8}{\sqrt{32}}cos2t, \frac{-8}{\sqrt{32}}sin2t, 0). \; and$$

$$\|T'(t)\| = \frac{8}{\sqrt{32}}. \; Then N = \frac{T'(t)}{\|T'(t)\|} = (-cos2t, -sin2t, 0).$$

Finally, to compute **k** we need to calculate the cross product of f', and f'' as follows:

$$f' \times f'' = \begin{bmatrix} i & j & k \\ -4sin2t & 4cos2t & 4 \\ -8cos2t & -8sin2t & 0 \end{bmatrix}$$

$$= i(32sin2t) - j(32cos2t) + k(32sin^22t + 32cos^22t).$$

Then,

$$\|f' \times f''\| = 32\sqrt{2}.$$

1.16   Exercise - 7

Find the length of the given curves on the specified interval:

1. $F(t) = (2t + 1, 4t, 1 - 2t)$, on the interval $1 \leq t \leq 2$.

2. $F(t) = 3ti + 2\cos t j + 2\sin t k$, on the interval $0 \leq t \leq \pi/2$.

3. $F(t) = cos6t, sin6t, 4t^{3/2}$, on the interval $0 \leq t \leq 1$.

Find the curvature **K**, unit normal vector **N**, and the unit tangent vector **T** for the given functions:

4. $f(t) = ti + 2t^2 j - k$.

5. $f(t) = (1 + \frac{1}{\sqrt{2}}t)i + (1 - \frac{1}{\sqrt{2}}t)j$.

6. $f(t) = 2ti + 1/t j$ *at* t=1.

7. $f(t) = t^3 j - k$ *at* $t = 0$.
See solutions on page-175.

Chapter 2

Functions of Several Variables

So far our attention was focused on functions with one variable which is real number called independent variable (input). The most common example for functions with one variable is the function of a straight line:

$$f(x) = mx + b. \qquad (2.1)$$

Where x is the single independent (input) real number, and y the single dependent (output) and real number too. The output can be scalar as in (1), or vector as in the vector valued functions discussed in the previous chapter.

Historical Notes:

Mary Fairfax Somerville (1780 - 1872)published her most well known book,"The Mechanics of Heavens" in 1831.
She was interested in problem of creating Geometric models for functions of several variables.

Examples of functions of single variable:

$$y \; = \; 3x + 1, \quad y = f(x).$$
$$z \; = \; 3t + 1, \quad z = f(t).$$
$$c \; = \; 5s + 7, \quad c = f(s).$$

Where the single independent variable in these three examples are: x, t, and s respectively.

2.1   Functions of 2 or 3 variables

In this section we will discuss functions with two ($\mathbf{R^2}$) or three ($\mathbf{R^3}$) variables. As done in the previous chapter we will examine the real valued functions of real point(*vector*) in $\mathbf{R^2}$and $\mathbf{R^3}$. The function of two variables is the rule that assign a unique value for each of pair (x, y) of numbers , for which the rule is defined. The function of two variables is written as:$z = f(x, y)$.

Graph Functions of 2-Variables:

Graphing a function of single variable $y = f(x)$ is an easy task, but graphing a function of two variables $y = f(x, y)$ is difficult, but frequent techniques are used to get a general idea of the graph of function with two variables. Since graphing a function with a single variable is easy, because of a single independent variable, then to graph a function with two independent variables will be possible if we set one of the two variables equal a constant, this will reduce the function into a single independent variable that can be graphed. For example: setting one of the independent variables $x = c$ in the function $z = f(x, y)$ gives $\Rightarrow z = f(c, y) = f(y)$ whose graph is the intersection of the desired graph with the plane $x = c$, similarly, if we set the other variable $y = c$ in the function $z = f(x, y)$

gives $\Rightarrow z = f(x,c) = f(x)$ whose graph is the intersection of the desired graph with the plane $y = c$ as a trace of f.

Example- 2.1:
Sketch several traces for the graph of the function : $z = x + y$.

Solution:
Tables(1), and (2) shows the equation of different traces.

Table(1) for $y = c$.

Plane y=c	Function z = f(x,c)
0	z= x
1	z= x + 1
2	z= x + 2
-1	z= x - 1
-2	z= x - 2

Plane y=c	Function z = f(x,c)
0	z= x
1	z= x + 1
2	z= x + 2
-1	z= x - 1
-2	z= x - 2

Table(2) for $x = c$.

The functions on both tables represents equations of straight lines that can be easily graphed.

Example- 2.2:
Sketch the level curves for the function $z = f(x,y) = x^2 + y^2$

Solution:
If we choose $z = r^2$ then $r^2 = x^2 + y^2$ represent curves of circles for $r > 0$, as explained in the table.

z	description of the curve
1	A circle with radius $= 1$
4	A circle with radius $= 2$
9	A circle with radius $= 3$

Example- 2.3:
Describe the graph of the function: $x^2 + y^2 + z^2 = 4$

Solution:
If we choose $y = z = 0$, the equation will be $x^2 = 4$, then $x = \pm 2$, this represent a circle of radius 2 in $xy - plane$. Then choosing $x = y = 0$ gives $z = \pm 2$, which represents a circle of radius 2 in $yz - plane$. Then the general picture represent a sphere centered at the origin and have a radius $r = 2$.

2.2   Exercise - 1

Sketch the following surfaces:

1. $x + y + z = 2$.

2. $3x - z = 1$.

3. $y = z^2$.

4. $\frac{x^2}{4} + \frac{y^2}{9} = 16$.

See solutions on page - 178.

2.3 Partial Derivatives

In this section we will use partial derivative and describe its affect on functions. The process of partial derivative can be described by answering the following question:

Question: How does the volume (V) in a cylinder vary with respect to its radius (r), and height (h) in the equation:

$$V = \pi r^2 h \tag{2.2}$$

Answer: We hold one of the two variables fixed, and study the other variable, then reverse the process in terms of the other variable being fixed, this precess is called "**Partial Differentiation**".

Note:
Partial differentiation ∂ is different from regular differentiation d

---

To differentiate a function of single variable $y = f(x)$ partially with respect to the independent variable x we write:

$$\frac{\partial y}{\partial x} = \frac{\partial f(x)}{\partial x}. \tag{2.3}$$

To differentiate a function of two - variables $z = f(x, y)$ partially with respect of the two independent variables x, and y, we have to do it in two steps with respect of the two variables separately:

1. Differentiation with respect to x:

$$\frac{\partial z}{\partial x} = \frac{\partial f(x, y\ fixed)}{\partial x}. \tag{2.4}$$

2. Differentiation with respect to y:

$$\frac{\partial z}{\partial y} = \frac{\partial f(x\, fixed, y)}{\partial y}.$$ (2.5)

Another symbol is used for partial differentiation as shown:

$$\frac{\partial z}{\partial x} = z_x.$$
$$\frac{\partial z}{\partial y} = z_y.$$

The formal definition of the differentiation is:

$$f_x(x, y) = lim_{\Delta x \to 0} \frac{f(x + \Delta x, y) - f(x, y)}{\Delta x}.$$ (2.6)

Application of Partial Differentiation:
Equations that involves partial differentiation are:
1. Heat Equations:

$$k\frac{\partial^2 u}{\partial x^2} = \frac{\partial u}{\partial t}, \quad k > 0$$ (2.7)

Where, $k = constant$, $u = velocity$, $x = space$, and $t = time$. Equation (2.7) is first order in time, and second order in space, it is used in the theory of heat flow in a rod, or thin wire.

2. Wave Equation:

$$k\frac{\partial^2 u}{\partial x^2} = \frac{\partial^2 u}{\partial t^2}, \quad k > 0$$ (2.8)

This equation is twice differentiable in both time and space. It is used in physics.

3. Laplace Equation:

$$\frac{\partial^2 u}{\partial x^2} + \frac{\partial^2 u}{\partial y^2} + \frac{\partial^2 u}{\partial z^2} = 0. \tag{2.9}$$

At steady-state equation(2.7) is independent of time, and hence reduces to equation (2.9) the Laplacian.

Example- 2.4:
Find $f_x(x, y)$, and $f_y(x, y)$ for the given function:
$f(x, y) = x^2 - y^2$.

Solution:

$$\frac{\partial f(x, y)}{\partial x} = f_x(x, y fixed) = 2x.$$
$$\frac{\partial f(x, y)}{\partial y} = f_y(x fixed, y) = -2y.$$

Example- 2.5:
Find $f_x(x, y)$, and $f_y(x, y)$ for the given function:
$f(x, y) = cos(2xy)$.

Solution:

$$\frac{\partial f(x, y)}{\partial x} = f_x(x, y fixed) = -2y sin(2xy).$$
$$\frac{\partial f(x, y)}{\partial y} = f_y(x fixed, y) = -2x sin(2xy).$$

Example- 2.6:
Find the value of $f_x(x, y)$, and $f_y(x, y)$ for the given function:
$f(x, y) = xy^2 - 4x^2$ at the point $(3, -2)$.

Solution:

$$\frac{\partial f(x,y)}{\partial x} = f_x(x, y\, fixed) = y^2 - 8x = -20.$$

$$\frac{\partial f(x,y)}{\partial y} = f_y(x\, fixed, y) = 2xy = -12.$$

The derivative of vector-function $r(t)$ is given as:

$$r'(t) = lim_{h \to 0} \frac{r(t+h) - r(t)}{h}. \qquad (2.10)$$

Provided that the limit exists. Also we can write :
If $r(t) = (x(t), y(t), z(t))$ and $x'(t)$, $y'(t)$, and $z'(t)$ exist, then,
$r'(t) = (x'(t), y'(t), z'(t))$.

Example- 2.7:
Find $r'(t)$ and find its values at $t = \pi/4$ for the vector function;
$r(t) = (1/2sint, 3/2cost)$.

Solution:

$$r'(t) = ((1/2sint)', (3/2cost)')$$
$$= (1/2cost, -3/2sint)$$

Then at $t = \frac{\pi}{4}$ gives $\Rightarrow$

$$r'(\pi/4) = (\frac{1}{2}cos(\frac{\pi}{4}), \frac{-3}{2}sin(\frac{\pi}{4}))$$
$$= (\frac{1}{2} \cdot \frac{1}{\sqrt{2}}, \frac{-3}{2} \cdot \frac{1}{\sqrt{2}})$$
$$= (\frac{1}{2\sqrt{2}}, \frac{1}{2\sqrt{2}}).$$

The Chain Rule:

To differentiate a function with a single(independent)variable :
$y = f(x)$ we write: $dy = f'(x)dx$. But if the independent variable x is a function of t, we write the function as $y = f(x(t))$ and to differentiate this we have to use the chain rule, and to do so, we will write the function as a composite then use the chain rule:

$$
\begin{aligned}
y &= f(g(t)) & (2.11) \\
&= (fog)(t), \; then \; differentiating \; gives, & (2.12) \\
y' &= f'(g(t)).g'(t) \; or, & (2.13) \\
\frac{dy}{dt} &= \frac{dy}{dx}.\frac{dx}{dt}. & (2.14)
\end{aligned}
$$

Equation (2.14) is called Leibniz notation.
If $z = f(x, y)$, and $x = x(t)$, $y = y(t)$,and both are differentiable then,

$$
\begin{aligned}
z &= f(x, y) & (2.15) \\
\frac{dz}{dt} &= \frac{\partial z}{\partial x}\frac{dx}{dt} + \frac{\partial z}{\partial y}\frac{dy}{dt} & (2.16)
\end{aligned}
$$

For a smooth function $z = f(x, y)$, function of two variables,

$$
\begin{aligned}
dz &= \frac{\partial z}{\partial x}dx + \frac{\partial z}{\partial y}dy, & (2.17) \\
dz &= z_x dx + z_y dy. & (2.18)
\end{aligned}
$$

dz is called the **total differential** of z.

And IF $z = f(x, y)$, $x = f(s, t)$, and $y = f(x, t)$, then this can be represented in tree diagram as follows:

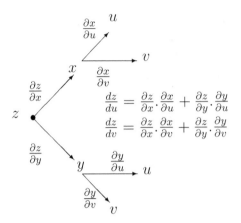

Now going back to the question at the beginning of this section:

Find the change in the volume of cylinder,

$$V = \pi r^2 h, \tag{2.19}$$

of radius r and height h when the radius is decreased by dr, and height increased by dh.

<u>Solution</u>: Applying equation (2.17) we can find the total differential of the volume ,

$$dV = \frac{\partial V}{\partial r} dr + \frac{\partial V}{\partial h} dh \tag{2.20}$$

$$= 2\pi r h \, dr + \pi r^2 \, dh. \tag{2.21}$$

Taking the initial volume to be V_0, the new volume V will be:

$$V = \pi (r - dr)^2 (h + dh). \tag{2.22}$$

Then we can compute the change in the volume as:

$$\Delta V = V - V_0$$

$$Or, \ \Delta V = \pi (r - dr)^2 (h + dh) - \pi r^2 h.$$

Example- 2.8: For the given cylinder, let $r = 2$, $h = 8$, $dr = -.1$, and
$dh = .2$, then $\Delta V = -7.54$ is the change in the volume of the cylinder.

For a smooth function with three independent variables, the total differential is:

$$w = f(x, y, z),$$
$$dw = \frac{\partial w}{\partial x} dx + \frac{\partial w}{\partial y} dy + \frac{\partial w}{\partial z} dz, \quad or$$
$$dw = w_x dx + w_y dy + w_z dz.$$

Example- 2.9:
Compute the total differential for the function:

$$u = x^2 + 2xy - y^3. \tag{2.23}$$

Solution:
Using equation (2.18) for two independent variables x, and y, gives,

$$du = u_x dx + u_y dy$$
$$= (3x^2 + 2y)dx + (2x - 3y^2)dy.$$

Differentiating Implicit Functions:

The Chain-Rule can be used to derive formula for Implicit Functions.
Let $F(x, y) = 0$ with $x = x(t)$, and $y = y(t)$, then $F(x(t), y(t)) = 0$ for all t in the domain. Then differentiating with respect to t,

$$\frac{d}{dt} F(x(t), y(t)) = \frac{\partial F}{\partial x} \frac{dx}{dt} + \frac{\partial F}{\partial y} \frac{dy}{dt}. \tag{2.24}$$

If $x(t) = t$, then the above equation becomes,

$$\frac{d}{dt}F\left(x(t), y(t)\right) = \frac{\partial F}{\partial x} + \frac{\partial F}{\partial y}\frac{dy}{dt} = 0. \qquad (2.25)$$

From this equation we get:

$$\frac{dy}{dt} = -\frac{\frac{\partial F}{\partial x}}{\frac{\partial F}{\partial y}} = -\frac{F_x}{F_y}, \quad f_y \neq 0. \qquad (2.26)$$

Velocity and Acceleration:

Now, we can combine our study of parametric equations, curves, and vector valued functions to form a model for the motion along a curve , which is used in physics.. In the previous chapter we found the directional vector to be:

$$r(t) = x(t)i + y(t)j. \qquad (2.27)$$

Where both x, and y are functions of t. Differentiating (2.27) with respect t gives the velocity:

$$Velocity \;=\; v(t) = \frac{dr(t)}{dt} = \frac{dx}{dt}i + \frac{dy}{dt}j, \, or \qquad (2.28)$$

$$v(t) \;=\; r'(t) = x'(t)i + y'(t)j. \qquad (2.29)$$

And differentiating (2.28) with respect to t gives the acceleration:

$$Acceleration \;=\; a(t) = \frac{dv}{dt} = \frac{dx'}{dt}i + \frac{dy'}{dt}j, \, or \qquad (2.30)$$

$$a(t) \;=\; v'(t) = r''(t) = x''(t)i + y''(t)j. \qquad (2.31)$$

Then, by taking the norm of the velocity in equation (2.) we can find the speed:

$$Speed = \|v(t)\| = \|r'(t)\| = \sqrt{(x'(t))^2 + (y'(t))^2}. \qquad (2.32)$$

Example- 2.10:
A time dependent directional function describes the motion of a spider in a circular path, is given as:

$$r(t) = 3sin(5/2t)i + 3cos(5/2t)j. \qquad (2.33)$$

Find the speed and acceleration of the spider in the path at any time t.

Solution:
The given equation has the following components:

$$x = 3sin(5/2)t$$
$$y = 3cos(5/2)t.$$

then, $x^2 + y^2 = (3sin(5/2)t)^2 + (3cos(5/2)t)^2 = 9$, which gives the radius of the path $= 3$.

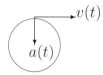

Taking any point on the path at any time the direction is:

$$
\begin{aligned}
Direction &= r(t) = 3sin(5/2t)i + 3cos(5/2t)j. \\
Velocity &= v(t) = r'(t) \\
&= (15/2)cos(5t/2)i - (15/2)sin(5t/2)j. \\
Acceleration &= a(t) = v'(t) \\
&= -(75/4)sin(5t/2)i - (75/4)cos(5t/2)j. \\
Speed &= \|v(t)\| \\
&= \sqrt{(15/2cos(5t/2))^2 + (15/2sin(5t/2))^2} \\
Then, \ Speed &= 15/2 = 7.5, speed \ of \ the \ spider.
\end{aligned}
$$

2.4   Exercise - 2

Find the partial differentiation for each of the following functions with respect to x, and y respectively:

1. $f(x,y) = \frac{2x-y}{x+y}$.

2. $f(x,y) = \sqrt{x^2 + y^2}$.

3. $f(x,y) = 6 - x + 4y$.

4. $f(x,y) = y\sqrt{x}$.

5. $f(x,y) = ln(x^2 - y^2)$.

6. $f(x,y) = e^{x^2 - y^2}$.

7. $f(x,y) = e^x cos(xy)$.

Use the limit definition from equation (2.6) to find $f_x(x,y)$, and $f_y(x,y)$ 8. $f(x,y) = x - 3y$.

9. $f(x,y) = x^2 + 2xy - y^2$.

10. $f(x,y) = \frac{4}{x+y}$.

Find the slopes of the surfaces in x, and y direction at the indicated point:

11. $f(x,y) = 5 - x^2 - y^2$, at $(2,2,3)$.

12. $f(x,y) = x^2 + y^2$, at $(3,1,-2)$.

13. $f(x,y) = e^{-y} siny$, at $(1,0,0)$.

Use Chain-Rule to find $\frac{dz}{dt}$ at the given t:

14. $z = x^2 + y$, $x = t^2 - 1$, $y = 2 + t^2$, at $t = 1$.

15. $z = y\cos xy$, $x = 1/t$, $y = 2t^3$, at $t = \pi/2$.

Find $\frac{\partial u}{\partial s}$, and $\frac{\partial u}{\partial t}$, for the following problems:

16. $u = x^3 - 3x^2y + y^2$, $x = s + 3t$, $y = 2s - t$.

17. $u = e^{y/x}$, $x = s + 2t$, $y = s - 2t$.

18. $u = e^y \cos y$, $y = s^2 - st - t^2$.

Compute the indicated partial derivative at the given values of s, and t: 19. $z = \frac{x^2+y^2}{2xy}$, $x = 3t - 5$, $y = t - 4s$, find $\frac{\partial z}{\partial s}$ at $t = 2$, and $s = -2$.

20. $u = x^2 + y^2$, $x = s\cos t$, $y = s\sin t$. Find $\frac{\partial u}{\partial s}$, and $\frac{\partial u}{\partial t}$, at $s = 3$, and $t = \pi/3$. See solutions on page-180.

2.5   Directional derivatives and the gradient

As done in the previous section, the partial derivative of function with two independent variables, $z = f(x, y)$, its components at x_0, and y_0 are:

$$z_x = f_x(x_0, y_0).$$
$$z_y = f_y(x_0, y_0).$$

Where, z_x, z_y measures the rate of change of f along the positive direction of the coordinate axes.

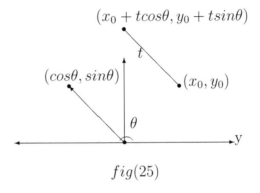

$$fig(25)$$

Then the directional derivative of f at (x_0, y_0) in the direction θ is the derivative of this function with respect to t evaluated at $t = 0$, and its called $f_\theta(x_0, y_0)$ and using the chain rule gives,

$$f_\theta = f_x cos\theta + f_y sin\theta. \tag{2.34}$$

Where the derivatives are evaluated at (x_0, y_0).

Definition: Directional Derivative:

If f is differentiable function of x, and y, then the derivative of f in the direction of the unit vector, $u = cos\theta i + sin\theta j$, is:

$$f_u = f_x cos\theta + f_y sin\theta. \qquad (2.35)$$

For fixed point $x = x_0 + t cos\theta, y = y_0 + sin\theta$.

The directional derivative for a function with three independent variables in R^3 is:

$$F_D = f_x cos\alpha + f_y cos\beta + f_z cos\gamma. \qquad (2.36)$$

Where, F_D is the directional function, and $\alpha, \beta, and \gamma$ are the angles between the vector and $x - axis$.

Example- 2.11:

Find the directional derivative for the function:

$$f(x, y, z) = xy^2 + yz + xz^2. \qquad (2.37)$$

At a point $p = (1, 2, 1)$, with direction $(1/\sqrt{2}, -1/\sqrt{2}, 2/\sqrt{2})$.

Solution:

Let the directional derivative be F_D. The given function has the following derivative components:

$$
\begin{aligned}
f_x &= y^2 + z^2, \quad cos\alpha = 1/\sqrt{2}. \\
f_y &= 2xy + z, \quad cos\beta = -1\sqrt{2} \\
f_z &= y + 2xz, \quad cos\gamma = 2/\sqrt{2}.
\end{aligned}
$$

Applying the formula (2.36), we get:

$$
\begin{aligned}
F_D &= f_x cos\alpha + f_y cos\beta + f_z cos\gamma. \\
&= (x^2 + z^2).(1/\sqrt{2}) + (2xy + z).(-1/\sqrt{2}) \\
&+ (y + 2xz).(2/\sqrt{2}). \\
&= 5/\sqrt{2} - 5/\sqrt{2} + 8/\sqrt{2} = 4\sqrt{2}.
\end{aligned}
$$

Example- 2.12:
Sketch the path of an object moving along the curve with directional function:

$$r(t) = 5t^2 i + t^3 j + 6tk, \ t \geq 0. \hspace{2cm} (2.38)$$

Find the velocity and acceleration of the particle at $t = 1$.

Solution:
The parametric components of $r(t)$ are:

$$
\begin{aligned}
x &= 5t, \\
y &= t^3, \\
z &= 6t.
\end{aligned}
$$

At time $t = 1$ equation (2.38) is: $r(t = 1) = 5i + j + 6k$, this means the particle moved from position $(0,0,0)$ to $(5,1,6)$ with velocity $v(1) = r'(1) = 10i + 3j + 6k$, and acceleration of $a(1) = r"(1) = 10i + 6tj$.

Example- 2.13:
For the function: $f(x,y) = x^2 + 1/2y^2 + 8$. Find the directional derivative at the point $(1,2)$ in the direction of the unit vector : $u = cos(\pi/6)i + sin(\pi/6)j$.

Solution:
The unit vector u, has a length $= \sqrt{cos^2(\pi/6) + sin^2(\pi/6)}$. Using formula (2.36), we will find the following values first: $f_x = 2x, \ f_y = 1, \ cos\pi/6 = \sqrt{3}/2, \ sin\pi/6 = 1/2$. Then at the given point $(1,2)$ the directional vector is:

$$
\begin{aligned}
F_D &= f_x cos\theta + f_y sin\theta. \\
&= 2x cos\theta + sin\theta. \\
&= 2(1).\sqrt{3}/2 + 1/2 \approx 2.23.
\end{aligned}
$$

Note:

If the length of the direction vector is not equal one-unit, then we have to normalize the given vector before applying the formula as shown in the next example:

Example- 2.14:

Find the directional derivative of the function:

$f(x, y) = 3y - 4xy + 5x$, at $p(1, 2)$, and $v = 4i + 3j$.

Solution:

First we have to normalize the vector to get u:

$$u = \frac{v}{\|v\|} = \frac{4i + 3j}{\sqrt{16 + 9}} = \frac{4i + 3j}{5} = \frac{4}{5} i + \frac{3}{5} j.$$
$$= \cos\theta\ i + \sin\theta\ j.$$

Then we find the following derivatives:

$$f_x = -4y + 5.$$
$$f_y = 3 - 4x.$$

Then, at point $(1, 2)$ the directional derivative is:

$$F_D(1, 2) = -12/5 - 3/5 = -3. \tag{2.39}$$

The Gradient:

The vector $f_z = f_x i + f_y j$ is known as the gradient vector, the standard symbol for the gradient is : grad f, or ∇f, or del f. The gradient ∇ itself is an operator that can be applied in 2-dimensions as:

$$\nabla = \frac{\partial}{\partial x} i + \frac{\partial}{\partial y} j. \tag{2.40}$$

$$Or,\ \nabla f = \frac{\partial f}{\partial x} i + \frac{\partial f}{\partial y} j.$$

Similarly, in 3-dimensions:

$$\nabla = \frac{\partial}{\partial x}i + \frac{\partial}{\partial y}j + \frac{\partial}{\partial z}k \qquad (2.41)$$

$$Or, \ \nabla f = \frac{\partial f}{\partial x}i + \frac{\partial f}{\partial y}j + \frac{\partial f}{\partial z}k$$

In directional derivative formula we had:

$$F_z = f_x cos\gamma + f_y sin\gamma. \qquad (2.42)$$

This can be written as : $F_z = (f_x, f_y).(cos\gamma, sin\gamma)$. But from the rules of Dot-Product between two vectors u_1, and u_2 with angle θ between them, we have:

$$u_1.u_2 = \|u_1\|\|u_2\|cos\theta. \qquad (2.43)$$

Since the norm $\|cos\gamma, sin\gamma\| = 1$ then we can write:

$$\begin{aligned} F_z \ &= \ (f_x, f_y).(cos\gamma, sin\gamma). \\ &= \ \|f_x, f_y\|\|cos\gamma, sin\gamma\|cos\theta. \\ Then, \ F_z \ &= \ \|f_x, f_y\|cos\theta. \end{aligned}$$

The last equation tells us that: $F_z = max$, when $cos\theta = 1 \Rightarrow \theta = 0$. $F_z = min$, when $cos\theta = -1 \Rightarrow \theta = \pi$, and $F_z = 0$, when $cos\theta = 0 \Rightarrow \theta = \pi/2$

Properties of the Gradient:

Let $f(x,y)$ be differentiable at the point $p(x,y)$, then:
1. If $\nabla f(x,y) = 0 \Rightarrow f_u(x,y) = 0$ for all u.
2. The direction of maximum increase of f is $\nabla f(x,y)$, and the maximum value of $f_u(x,y) = \|\nabla f(x,y)\|$.
3. The direction minimum increase of f is $-\nabla f(x,y)$., and the

minimum value of $f_u(x, y) = -||\nabla f(x, y)||$.

Example- 2.15:
Find the maximum value of the given function:
$f(x, y) = 18 - 3x^2 - y^2$, from the point $p = (2, -3)$, and find the rate of increase.

Solution:
Using the gradient formula:

$$
\begin{aligned}
\nabla f(x, y) &= f_x i + f_y j. \\
\nabla(18 - 3x^2 - y^2) &= -6xi - 2yj. \\
Then\ at\ pint\ (2, -3)\ , \nabla(2, -3) &= -12i + 6j.
\end{aligned}
$$

And the rate of increase is : $||\nabla f|| = \sqrt{(-12)^2 + (6)^2} = 6\sqrt{5}$.

2.6   Exercise - 3

Find the gradient of the given functions:
1. $f(x, y) = x^2 + y^3 + 2x - y^2$.

2. $f(x, y) = y\cos x$.

3. $f(x, y) = 5x - 2y$.

Find the gradient of the given functions at the indicated point:
4. $f(x, y) = 10 - 3x + 5y^2, p(2, 1)$

5. $f(x, y) = \cos(x + y), p(2, 0)$.

6. $f(x, y) = \ln(x - y^2), p(3, 2)$. See solutions on page - 184.

2.7 Constrained Optimization: Lagrange Multiplier

Constrained Optimization means constrained minima, and maxima. Finding maximum, and minimum for a function with two variables can be done by taking the gradients and applying second derivative test. Sometimes the problem involves some side-conditions(constrained), in such cases we have to apply the method of Lagrange Multiplier. The following are examples of these cases:

1. Find the shortest distance from a point $p(x_0, y_0)$ to a line $y = mx + b$.

To use the Lagrange Method, this question can be written as follows:

Maximize the function: $z = \sqrt{(x - x_0)^2 + (y - y_l)^2}$,
subject to the constraints: $y - mx - b = 0$.

2. Maximize the area that can be inscribed in the ellipse: $x^2 + 4y^2 = 4$.

This question can be written as follows:

If (x,y) is the corner point of the rectangle in the first quadrant with sides $2x, 2y$.

Maximize the function: $A = f(x, y) = 4xy$.

Subject to the constraints: $g(x, y) = x^2 + 4y^2 - 4 = 0$.

The method of Lagrange Multiplier was discovered by the French-Italian Mathematician Joseph Louis Lagrange (1736-1813).

Description of Lagrange Method:

To describe the method of Lagrange Multiplier, we will consider functions with one and more variables:

1. For a function with one independent variable : $y = f(x)$:
Find $\nabla f(x) = \lambda \nabla g(x)$. with the constrained $g(x) = 0$.

2. For a function with two independent variables:
Find $\nabla f(x, y) = \lambda \nabla g(x, y)$. with the constrained $g(x, y) = 0$.

3. For a function with three independent variables :
$y = f(x, y, z)$: Find $\nabla f(x, y, z) = \lambda \nabla g(x, y, z)$. with the constrained $g(x, y, z) = 0$.

Note: The symbol λ is the Lagrange Multiplier.

Example- 2.16:

Find the extreme values of the function of two independent variables x, and y: $f(x, y) = x^2 y$ that is subject to the constraints : $g(x, y) = x_2 + y^2 - 4 = 0$.

Solution:

To apply the method, we write:

$$\begin{aligned}
\nabla f(x, y) &= \lambda \nabla g(x, y). \\
\nabla (x^2 y) &= \lambda \nabla (x^2 + y^2 - 4 = 0). \\
(f_x, f_y) &= \lambda (g_x, g_y). \\
(2xy, x^2) &= \lambda (2x, 2y).
\end{aligned}$$

From the last equation we get:

$$2xy = 2x\lambda. \qquad (2.44)$$
$$x^2 = 2y\lambda. \qquad (2.45)$$
$$And, \ x^2 + y^2 - 4 = 0 \qquad (2.46)$$

Solving the system of 3-equations: Equation (2.44) gives:
$2xy - 2x\lambda = 0 \Rightarrow 2x(y - \lambda) = 0$ then, $\Rightarrow x = 0, or \ y = \lambda$.

For $x = 0 \Rightarrow$ then equation (2.46) gives $y = \pm 2$, and equation
(2.44) gives: $x^2 = 2y^2$, $\Rightarrow$ $x = \pm\sqrt{2}y$.
Substituting $x = \pm\sqrt{2}y$ in equation (2.46) gives $y = \pm 2/\sqrt{3}$,
substituting back to get $x = \pm 2\sqrt{2/3}$.
Thus the points are:

(x,y) points	$f(x,y)$ values
$(0,2)$	0
$(0,-2)$	0
$(2\sqrt{2/3}, 2/\sqrt{3})$	$16/3\ \sqrt{3}$
$(-2\sqrt{2/3}, 2/\sqrt{3})$	$16/3\ \sqrt{3}$
$(2\sqrt{2/3}, -2/\sqrt{3})$	$-16/3\ \sqrt{3}$
$(-2\sqrt{2/3}, -2/\sqrt{3})$	$-16/3\ \sqrt{3}$

Thus the table shows a maximum value for the function
f(x,y)$= \pm 16/3\sqrt{3}$, at the points $(\pm 2\sqrt{2/3}, 2/\sqrt{3})$,
and minimum values at the points $(\pm 2\sqrt{2/3}, -2/\sqrt{3})$.

Example- 2.17:
Find the points on the sphere $x^2 + y^2 + z^2 = 4$, at maximum,
and minimum distance from the point: $p(1, 1, 2)$.

Solution:
The question means Minimize, and maximize the given
function: $f(x, y) = (x - 1)^2 + (y - 1)^2 + (z - 2)^2$.
subject to the constraints: $g(x, y, z) = x^2 + y^2 + z^2 - 4 = 0$.
Applying the formula and the steps as in the previous example:

$$
\begin{aligned}
\nabla f(x, y, z) &= \lambda \nabla g(x, y, z). \\
\nabla((x - 1)^2 + (y - 1)^2 + (z - 2)^2) &= \lambda \nabla (x^2 + y^2 + z^2 - 4 = 0). \\
(f_x, f_y, f_z) &= \lambda(g_x, g_y, g_z). \\
(2(x - 1), 2(y - 1), 2(z - 1)) &= \lambda(2x, 2y, 2z).
\end{aligned}
$$

From the last equation we get:

$$2(x - 1) = 2x\lambda. \tag{2.47}$$
$$2(y - 1) = 2y\lambda. \tag{2.48}$$
$$2(z - 2) = 2z\lambda. \tag{2.49}$$
$$And, \quad x^2 + y^2 + z^2 - 4 = 0 \tag{2.50}$$

Solving the system of 4-equations:
First solving equations (2.47, 48, 49) for $x, y,,$ and z in terms of λ gives:

$$x = \frac{1}{1 - \lambda}; y = \frac{1}{1 - \lambda}; z = \frac{2}{1 - \lambda}. \tag{2.51}$$

Then substituting these values of x, $y,,$ and z into equation (2.50) to find the value of λ gives:

$$\lambda = 1 \pm \sqrt{\frac{3}{2}}. \tag{2.52}$$

Substituting the value of λ back into the above 3- equations to get the following numerical values for x, y, and z:

$$x = \frac{\sqrt{6}}{3}; y = \frac{\sqrt{6}}{3}; z = \frac{2\sqrt{6}}{3}. \tag{2.53}$$

The theory of optimization can be applied to a variety of practical problems in real life, such in the following example:

Example- 2.18:

A house owner wishes to fence his rectangular shaped backyard. He bought 1500 yards of fence from home-depot. What are the dimension of the maximum area that he can enclose with this fence.

Solution:

Since the backyard is in a rectangular shape: Then it has a perimeter $P = 2x + 2y = 1500$. Then we can state the optimization problem here as:

$$Maximize: \quad A \quad = \quad xy. \quad (2.54)$$
$$Subject\ to\ the\ constrained: 2x + 2y \quad = \quad 1500. \quad (2.55)$$

First we will express equation (2.55) in terms of a single variable y as:

$$2y \quad = \quad 1500 - 2x. \quad\quad (2.56)$$
$$y \quad = \quad 750 - x. \quad\quad\quad (2.57)$$

Substitute (2.57) into (2.54):

$$A = x(750 - x) = 750x - x^2. \quad\quad (2.58)$$

Equation (2.58) is a second order equation , and is a parabola opened down as shown in the graph:

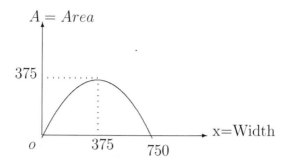

$fig(26)$ $represents$ $graph$ of $A = x(750 - x)$

The domain of the acceptable positive values for x (width) are: $0 < x < 750$.

It is clear from the graph that the maximum value of x is $1/2$ a distance between $(0, 750)$, or $x \simeq 375$ yards. By substituting in equation (2.58) we get the value of $y = 375$ Then the dimensions are: $x = y = 375$. this means the rectangle of perimeter of 1500 yards having a maximum area is a square of 375×375 yards.

2.8 Exercise - 4

1. Using Lagrange Multiplier, minimize the volume $V = xyz$ with constraint: $g(x, y, z) = 3xy + yz + xz - c$. c is a constant.

2. In business, a company modeled its profit in 3-productions x, y, z as: $p(x, y, z) = 2x + 6y + 4z$ with manufacturing constraint force $:x^2 + 2y^2 + z^2 \leq 600$. Find the maximum profit for the company.

3. To construct an open-top-rectangular box with fixed volume V. Minimize the amount of wood needed. See solutions on page - 184.

Chapter 3

Multiple Integrals

3.1　Single Integral

First we will try to illustrate the idea of integration as follows:
Let f be a non-negative, and continuous function in the closed
interval $[a, b]$, or $a \leq x \leq b$.

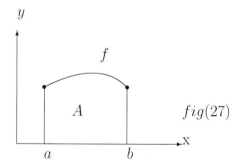

$fig(27)$

To find the area under the region that is bounded by the
curve f, and the two vertical lines at a, and b , we use the

integral as follow:

$$Area = A = \int_a^b f(x)dx. \qquad (3.1)$$

Where, the integral over the interval $[a, b]$ represents the length, and $f(x)$ represents the height.

Using the rule of the fundamental theorem from Calculus, the integral on (3.1) can be solved as follows:

$$\int_a^b f(x)dx = F(b) - F(a). \qquad (3.2)$$

3.2   Method of Strips for area under the curve

Since the double integrals used to find the area under the curve has been covered in Calculus-II, we are not covering it in this book, but we will just mention this unique method that is used to find the area under the curve.

1. If the region R is defined by $a \leq x \leq b$, and $f_1(x) \leq y \leq f_2(x)$, whee $f_1(x)$, $f_2(x)$ are continuous on the closed interval $[a, b]$, then the area of R is given by, and shown in the fig(28) :

$$A = \int_a^b \int_{f_1(x)}^{f_2(x)} dy \; dx. \qquad (3.3)$$

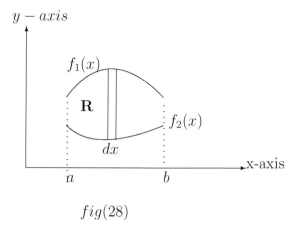

$fig(28)$

2. If the region R is defined by $c \leq y \leq d$, and $f_1(y) \leq x \leq f_2(y)$, whee $f_1(y)$, $f_2(y)$ are continuous on the closed interval $[c, d]$, then the area of R is given by, and shown in the fig(29):

$$A = \int_c^d \int_{f_1(y)}^{f_2(y)} dx \, dy. \tag{3.4}$$

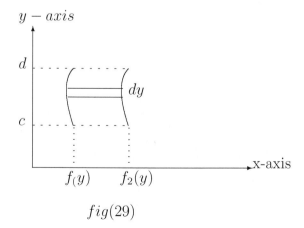

$fig(29)$

Example- 3.1:

Find the area of the region bounded between : $y = 2x$, and $y = x^2$ shown in the fig(30), using strip method.

Solution:

The two curves gives us the points of intersection as:

$$y = x^2 \ = \ 2x.$$
$$x^2 - 2x \ = \ 0. Solving \ for \ x, \ gives :$$
$$x \ = \ 0, \Rightarrow x = 2. Then solving for y gives :$$
$$for \ x \ = \ 0 \to y = 0,$$
$$for \ x \ = \ 2 \to y = 4.$$

Then the two points of intersection s are: $p_1 = (0,0)$, and $p_2 = (2,4)$.

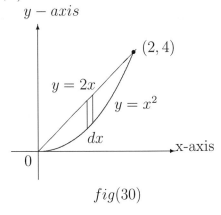

$$fig(30)$$

Then the area between the two curves is : (top curve - bottom curve). Δx, which is equivalent to the integral over the region, and between $x = 0$, to $x = 2$:

$$A \ = \ (top \ curve - bottom \ curve).\Delta x.$$
$$= \ (2x - x^2)\Delta x.$$
$$= \ \int_0^2 (2x - x^2)dx.$$

$$A = x^2 - x^3/3]_0^2 = 1/2.$$

The same result can be achieved if we take the strip on the horizontal position.

3.3 Double Integrals

As seen in the previous section, the single integral applied to a function with single independent variable represents the area. The double integral applied to a function with two independent variables represents the Volume.

Definition:
Consider a continuous function $f(x, y)$ of two independent variables x, and y, then the region bonded by this function can be described as a double integral written as follows:

$$\int\int_R f(x, y) dA \qquad (3.5)$$

Where, R is the region of integration. Now we will show how this integral represents the volume of the region.

Example- 3.2:

Consider a box with the following dimensions:
Length = [0,a] on x-axis.
width = [c,d] on y-axis.
Height = [0,e] on z-axis.

Geometrically, the volume of the box = Area of the base x the height. In integral form,

$$Area\ of\ the\ base\quad =\quad [c,a] \times [c,d].$$

$$=\quad \int_c^d \int_c^a dxdy.$$

$$=\quad \int\int_R dA.$$

And the height of the box is represented by the function $f(x,y)$, then the volume in integral form is:

$$V = \int_c^d \int_0^a f(x,y)dxdy \quad = \quad \int\int_R f dA$$

$$=\quad Area\ of\ the\ base\ \times the\ height.$$

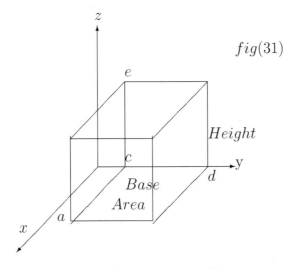

$fig(31)$

And using triple integrals will be;

$$V = \int_0^e \int_c^d \int_0^a f(x,y)dxdydz \quad = \quad \int\int_R f dA$$

$$=\quad Length \times height \times width.$$

3.4 Finding the volume of a box
in two different approaches

Example-(3.3):

Let $f(x,y) = 2x^2 + xy + y^2$ be a function with two independent variables, and on the box bounded by the region: $\mathbf{R} = [0,2] \times [0,1]$. Find the volume of the box.

First Approach: Using single integrals one at a time:

We will take the first single integral on the y-intervals:

$$
\begin{aligned}
\int_0^1 f(x,y)dy &= \int_0^1 (2x^2 + xy + y^2)dy. \\
&= \int_0^1 2x^2 dy + \int_0^1 xy dy + \int_0^1 y^2 dy. \\
&= [2x^2 y + xy^2/2 + y^3/3]_0^1. \\
&= 2x^2 + 1/2x + 1/3.
\end{aligned}
$$

Now we take a second single integral over the x-interval with f(x,y) = the result of the first integral.

$$
\begin{aligned}
\int_0^2 (2x^2 + 1/2x + 1/3)dx &= \int_0^2 2x^2 dx + \int_0^2 1/2x + \int_0^2 1/3)dx' \\
&= 2x^3/3 + x^2/4 + 1/3x]_0^2. \\
&= 7
\end{aligned}
$$

Second Approach: Taking two integrals together:

Here we will write the double integrals, and integrate the first one from inside going out:

$$
\begin{aligned}
V = \int_0^1 \int_0^2 f(x,y)dxdy &= \int_0^1 \int_0^2 (2x^2 + xy + y^2)dxdy. \\
&= \int_0^1 \{\int_0^2 2x^2 dx + \int_0^2 xy dx \\
&+ \int_0^2 y^2 dx\}dy.
\end{aligned}
$$

$$V = \int_0^1 [2x^3/3 + yx^2/2 + y^2 x]_0^2 dy.$$
$$= 16/3y + y^2 + 2/3y^3]_0^1.$$
$$Then, \ V = 7$$

Notice from this example, how changing the order of integrand did not change the result.

The double integrals as an Iterated integral:

Suppose f is a continuous function on a closed box with intervals: $k = [a, b] \times [c, d]$ then the double integral of f over k and the two iterated integrals are equal.

$$\int \int_k f(x, y) dA = \int_a^b \int_c^d f(x, y) dy dx = \int_c^d \int_a^b f(x, y) dx dy.$$
$$(3.6)$$

This was seen in the above example, and proves that double integral is expressible as an **iterated integral** and that the corresponding relation holds for the single integrals.

Equality of Mixed Partials:

Let f be a continuous function located in a box that is bounded by $[a, b] \times [c, d]$ for $a \le x \le b$, and $c \le y \le d$. Consider the sub box bounded by $[a, x] \times [x, y]$, as shown in $fig(32)$, then the double integral of f over this sub box is a function of x, and y and we can express it as an iterated integral as follows:

$$G(x, y) = \int \int_{[a,x] \times [c,y]} f.$$
$$= \int_a^x \int_c^y f(x, y) \ dt \ ds.$$
$$= \int_c^y \int_a^x f(x, y) \ ds \ dt.$$

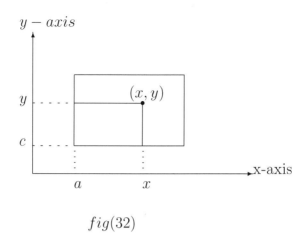

$fig(32)$

3.5 Exercise - 1

Integrate the following double integrals:

1. $\int_0^1 \int_1^2 (2x^2 + y^2 + xy + 4) dy dx$.

2. $\int_0^1 \int_0^1 x \; e^{x-y} \; dy \; dx$. See solutions on page- 187.

3.6 Integration over Standard Regions

As seen in the case of the box above, integrating in different order is possible, but if double integral is applied to functions defined over a region more general than closed box called"standard regions", then complications will arise if we try to iterate the integrals in the other order. But it can be done if we choose the order of integration with the consideration of the continuity of the limits of the last integration.

Suppose f is a continuous function in a region R , $fig(33)$.

Type-I: For each fixed x, the integral gives:

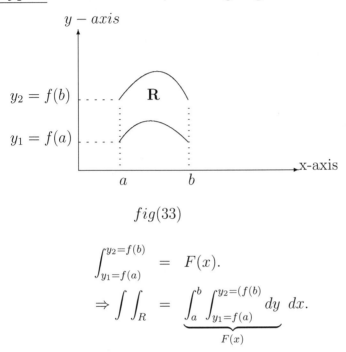

$$fig(33)$$

$$\int_{y_1=f(a)}^{y_2=f(b)} = F(x).$$

$$\Rightarrow \int\int_R = \underbrace{\int_a^b \int_{y_1=f(a)}^{y_2=(f(b)} dy \ dx.}_{F(x)}$$

Complications will arise if we try to iterate the integrals in the other order.

Type-II:

Similarly, if f is a continuous function on the region R-$fig(34)$:

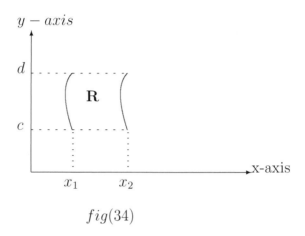

$$fig(34)$$

With y fixed in the interval $[c, d]$, the integral for fixed y is;

$$\int_{x_1=f(c)}^{x_2=f(d)} = F(y).$$

$$\Rightarrow \int\int_R = \underbrace{\int_c^d \int_{x_1=f(c)}^{x_2=(f(d)} dx \; dy.}_{F(y)}$$

But we do not attempt to interchange the other order.

Note: The development of the double integral over the standard region can be reduced to the case of the closed box, if we take it in this way:

$x - intervals = [a, b]$.

$y - intervals = [min.c, max.d]$.

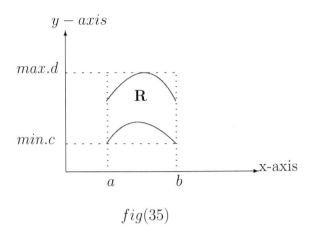

$$fig(35)$$

The box is $\Rightarrow \mathbf{B_R} = [a, b] \times [\, min.c, max.d\,]$.

If the region of integration R is neither of type-I, nor Type-II, it may be possible to divide the region into two or more sub-regions, each of which is either of Type-I, or type-II, as shown in the fig(36).

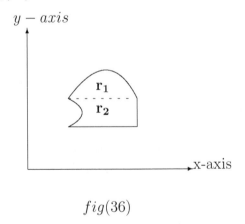

$$fig(36)$$

Then the region in fig(36) can be found by taking the sum of the two subregions: $\int \int_R f(x, y)dA = \int \int_{r_1} f(x, y)dA + \int \int_{r_2} f(x, y)dA$.

3.7 Exercise - 2

Evaluate the Iterated Integrals:
1. $\int_0^1 \int_0^\pi x \, siny \, dy \, dx$.

2. Find the volume of the region bounded on top by : $z = 3x + y + 2$, on the bottom by the $xy - plane$, on the sides by the planes:
$x = 0, x = 2, and \, y = 1, y = 2$.

3. Find the volume of the region bounded by the graph of $f(x, y) = 3x^2 + y^2 sin\pi x$, the $xy - plane$, and the planes:
$x = 0, x = 2, and \, y = -2, y = 2$. See solutions on page-188.

3.8 Double Integrals Over General Regions in The Plane

If the integral is applied over an arbitrary bounded region D in the plane, we say that D is an elementary region if it can be described as a subset of R^2 of one of the following 3-types:
Type-I:$D = \{(x, y) | \gamma \leq y \leq \delta, \; a \leq x \leq b|\}$.

Type-II:$D = \{(x, y) | \beta \leq y \leq \alpha, \; c \leq x \leq d|\}$.

Type-III: D is of both Type-I, and Type-II.

Note: Type-I region D has boundary consisting of straight segments on left, and right, and graph of continuous function of x on top, and bottom.
Type-II region D has boundary consisting of straight segments on top, and bottom, and graph of continuous functions of y on left, and right.

Example-(3.4):

Two concentric curves, a circle of radius 1, and an oval of radius 2, as shown in fig(37). Evaluate the double integral as the sum of integrand over the subregions.

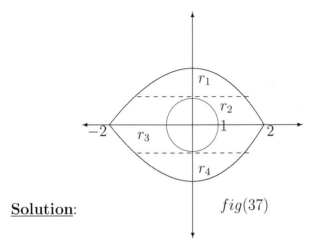

Solution: $fig(37)$

The integral of the region between the two closed curves is computed by taking the sum of the double integral over each subregion:

$$\int\int_R = \int\int_{r_1} f dA + \int\int_{r_2} f dA + \int\int_{r_3} f dA + \int\int_{r_4} f dA. \quad (3.7)$$

3.9   Triple Integrals

As done before in section (3.4), to find the volume of the box we found:

$$V = \int \int_R f(x,y)dA = area\ of\ the\ base \times height. \quad (3.8)$$

Here we will evaluate the volume of the box by taking triple integrals over the three intervals of the sides of the box, $a \leq x \leq b$, $c \leq y \leq d$, and $e \leq z \leq f$, and the integrand can be written in 6-different orders as follows:

$$\begin{aligned}
V &= \int \int \int f\,dv. \\
&= \int \int \int f(x,y,z)dx\,dy\,dz. \\
&= \int \int \int f(x,y,z)dx\,dz\,dy. \\
&= \int \int \int f(x,y,z)dy\,dx\,dz. \\
&= \int \int \int f(x,y,z)dy\,dz\,dx. \\
&= \int \int \int f(x,y,z)dz\,dx\,dy. \\
&= \int \int \int f(x,y,z)dz\,dy\,dx.
\end{aligned}$$

Fubini's Theorem:

Fubini in his theorem has related, the triple integral of a continuous function f bonded on closed intervals : $[a, b] \times [c, d] \times [e, f]$, to all the six orders shown above, and as done in the Tree-diagram shown below:

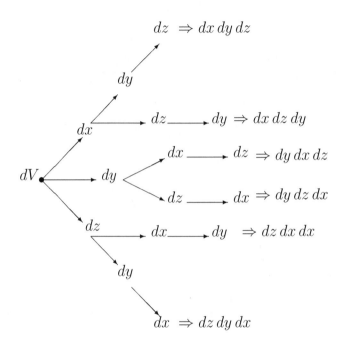

Finding intervals in multiple integration:

Example- (3.5):
Find the volume of the tetrahedron formed by $x + \frac{y}{2} + \frac{z}{4} = 1$, and the planes: $x = 0, y = 0$.

Solution:
1. we need to graph the tetrahedron by finding points, and connect them:

$$\text{Let, } y = z = 0 \Rightarrow x = 1 \Rightarrow Pt = (1,0,0).$$
$$y = z = 0 \Rightarrow x = 1 \Rightarrow Pt = (0,2,0).$$
$$y, z = z = 0 \Rightarrow x = 1 \Rightarrow Pt = (0,0,4).$$

2. We find the intervals on the axes: Where,

$$I = \underbrace{\int \int \int f(x,y,z) dx dy dz.}_{R}$$

$$= \int_{a_1}^{a_2} \int_{g_1(x)}^{g_2(x)} \int_{h_1(x,y)}^{h_2(x,y)} f(x,y,z) dz dy dx.$$

$I_y = \longrightarrow$ set $x = z = 0$, and solve for y.
Then, $y/2 = 1 \rightarrow y = 2$.
$I_z = f(y)$, set $x = 0$ and solve for z . $z = 4 - 2y$.

$I_x = f(x, z)$, solve for x. $x = 1 - y/2 - z/4$.
Then the intervals are:

$$
\begin{aligned}
I_y &= [0, 2]. \\
I_z &= [0, (4 - 2y)]. \\
I_x &= [0, (1 - y/2 - z/4)].
\end{aligned}
$$

Then the volume integrals are:

$$
\begin{aligned}
V &= \int_0^2 \int_0^{(4-2y)} \int_0^{(1-y/2-z/4)} dx\,dz\,dy. \\
&= \int_0^2 \int_0^{(4-2y)} [z]_0^{(1-y/2-z/4)} dz\,dy. \\
&= \int_0^2 \int_0^{(4-2y)} (1 - y/2 - z/4) dz\,dy. \\
&= \int_0^2 [z - yz/2 - z^2/8]_0^{(4-2y))} dx. \\
&= 2 \int_0^2 (1 - y/2) dy. \\
&= -4/3(1 - y/2]_0^2. \\
Then,\ V &= 4/3.
\end{aligned}
$$

Example-(3.6):

The function of three independent variables: $f(x, y, z) = xe^y + 2xyz$ is bounded by : $[-1, 2] \times [0, 1] \times [0, 3]$. Find the volume of the bounded region.

Solution:

Function f is continuous and hence satisfies Fubini's Theorem. Therefore:

$$
\int \int \int (xe^y + 2xyz) dv = \int_0^3 \int_0^1 \int_{-1}^2 (xe^y + 2xyz) dx\,dy\,dz.
$$

$$= \int_0^3 \int_0^1 \{ \int_{-1}^2 (xe^y + 2xyz) \} dydz.$$

$$= \int_0^3 \{ \int_0^1 (3/2(e-1) + 3/2z)) \} dz.$$

$$= 3/2(e-1)z + 3/4z^2 \ ]_0^3.$$

$$= 41.54.$$

Example-(3.7):
Find the volume of the solid bounded by the coordinate planes
and the plane: $x + 2y + 3z = 6$.

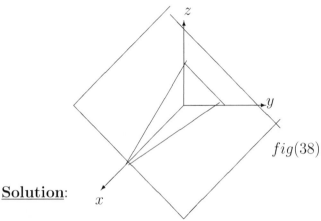

$fig(38)$

Solution:

To graph the given linear equation of the plane, we found its
points of intersections with the 3-axes to be: $(6, 0, 0)$, $(0, 3, 0)$,
and $(0, 0, 6)$ which indicates that these are the vertices of tri-
angular shape, as shown in the fig(38).
Using one variable in terms of the other two independent variables, we can write:

$$z = f(x, y) = \frac{6 - x - 2y}{3.} \qquad (3.9)$$

Then the volume of the solid can be found:

$$V = \int \int f(x, y) dA.$$

$$= \int_0^3 \int_0^{6-2y} \frac{6-x-2y}{3} dxdy.$$

$$V = \cdots \int \frac{6-x-2y}{3} dx \cdots.$$

$$= 1/2 \int_0^3 (12 - 8y + y^2)dy = 9/2.$$

3.10   Exercise - 3

Use the triple integral to verify that the volume of a sphere of radius r is $\frac{4}{3}\pi r^3$. See solution on page-189.

3.11   Transformation of Variables

Double Integrals in Polar Coordinates:

In solving integral problems, it is more convenient (easier) sometime, to evaluate the region in $xy - plane$ in polar coordinates. For example: the region in the first quadrant in fig(39) bounded by , $y = x$, and the curve of $x^2 + y^2 = 4$, $\theta = \pi/4$.

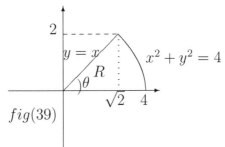

$fig(39)$

In this problem it is easier to use polar coordinates than rectangular one.

The rectangular coordinates, and the domain for the problem are:

$$x = rcos\theta.$$
$$y = rsin\theta.$$
$$R = \{(r,\theta) : 0 \le r \le \theta, 0 \le \theta \le \pi/4\}.$$

Note: Changing integration to polar coordinates is desirable if carrying out the integration in rectangular is more difficult.

The double integral of the function $f(x,y)$ over the region R as shown in fig(39) is:

$$\int\int_R f(x,y)dA = \int_y \int_x f(x,y)dxdy. \qquad (3.10)$$

With intervals and the function as:

$$x - intervals \Rightarrow x = 0 \text{ to } x = \sqrt{4 - y^2}.$$
$$y - intervals \Rightarrow y = 0 \text{ to } y = \sqrt{2}.$$
$$f(x,y) \Rightarrow F(r,\theta) : (rcos\theta)^2 + (rsin\theta)^2 = 4.$$
$$\Rightarrow r = 2.$$

Hence the double integral is:

$$\int\int_R f(x,y)dA = \int_0^2 \int_y^{4-y^2} f(x,y)dxdy.$$
$$= \int_0^{\pi/4} \int_0^2 f(r,\theta)rdrd\theta.$$

To evaluate the integral, $F(r,\theta)$ represents the volume above R and under the surface $z = f(r,\theta)$.

Triple Integrals in cylindrical Coordinates:

The components in cylindrical coordinates are:

$$x = rcos\theta.$$
$$y = rsin\theta.$$
$$z = z.$$
$$f(x, y, z) = (rcos\theta, rsin\theta, z).$$

Thus the triple integral in cylindrical coordinates is:

$$V = \int\int\int f(x, y, z)dv.$$
$$= \int\int\int f(r, \theta, z)rdrd\theta dz.$$

Example-(3.8):

Evaluate the triple integral ,
$\int\int\int_R (y^2 x + x^2 z)dv$, over the cylindrical coordinates:
$(r, \theta, x); 1 \leq r \leq 2, 0 \leq \theta \leq \pi, 0 \leq x \leq 3$.

Solution: The triple integration for the problem is:

$$\int\int\int_R (y^2 z + x^2 z)dv = \int\int\int z\underbrace{(y^2 + x^2)}_{r^2}.rdrd\theta dz.$$
$$= \int_0^3 \int_0^\pi \int_1^2 zr^2.rdrd\theta dz.$$
$$= \cdots \int_1^2 zr^3 dr \cdots.$$
$$= \cdots zr^4/4]_1^2 \cdots.$$
$$= \cdots \int_0^\pi (15/4z)d\theta \cdots.$$

$$= \int (15/4z\pi)dz.$$
$$= 15/4\pi z^2/2]_0^3.$$
$$= 138/8\pi.$$

Triple Integrals in Spherical Coordinates:

The triple integral in spherical coordinates is:

$$\int \int \int_R f(x,y,z)dv = \int \int \int f(\rho,\theta,\phi)\rho^2 \, sin\phi \, d\rho \, d\theta \, d\phi.$$

Example-(3.9):

Evaluate $\int \int \int (x^2 + y^2 + z^2)^{3/2}dv$, with the domain:
$x \geq 0, y \geq 0, z \geq 0$, using spherical coordinates.
Where $\{0 \leq \rho \leq 2, 0 \leq \theta \leq \pi/3, 0 \leq \phi \leq \pi/3\}$.

Solution:

The triple integral for the question, in spherical coordinates is:

$$\int \int \int_R f(\rho,\theta,\phi)\rho^2 \, sin\phi \, d\rho \, d\theta \, d\phi$$
$$= \int \int \int (x^2 + y^2 + z^2)^{3/2}dv.$$
$$= \int_0^{\pi/3} \int_0^{\pi/3} \int_0^2 (\rho^2)^{3/2}\rho^2 \, sin\phi \, d\rho \, d\theta \, d\phi.$$
$$= \cdots \int_0^2 \rho^5 d\rho \cdots.$$
$$= 32/3 \int_0^{\pi/3} \int_0^{\pi/3} sin\phi d\theta d\phi.$$
$$= 16\pi/9.$$

3.12   Exercise - 4

1. Using the triple integrals , find the volume of the ellipsoid:

$$\frac{x^2}{4} + \frac{y^2}{9} + \frac{z^2}{16} = 1.$$

2. Find the volume of the solid in the first octant bounded by the cylinder $x^2 + z^2 = 9$, the plane $x + y = 4$, and the three coordinates.

3. Using triple integrals in cylindrical coordinates, find the volume of the region inside the sphere: $x^2 + y^2 + z^2 = 9$, and the cylinder $(x - 1)^2 + y^2 = 1$.

4. Using the spherical coordinates, find the volume of the solid inside the sphere:$x^2 + y^2 + z^2 = 4$ and outside the cone : $z^2 = x^2 + y^2$.

See solutions on page 189.

Chapter 4

Line and Surface Integrals

4.1 Line Integrals

Up to now we have studied the following integrals:

1. Single Integrals integrated over $I = [a, b]$, given as $\int_a^b f(x)dx$.

2. Double integrals over the region R of two independent variables x, y.

Now we will study a new type of integral, that is integrated over a curve called" Line Integral", and written as:

$$\int_c f(x, y)ds. \tag{4.1}$$

For which we integrate over a piecewise smooth curve C. This type of integral is used in studying properties of vector field, and many physical applications. So in this type of integral, instead of integrating over a given and closed interval, we will be integrating over a curve in the space, with a given positive path. equation (4.1) sometimes is written as:

$$\oint_c f(x,y)ds. \tag{4.2}$$

Where the symbol $\oint$ indicates calculating line integral. The line integral for a scalar field f with respect to arc length is:

$$\oint_c fds = \int_a^b f(r(t))(\frac{ds}{dt})dt. \tag{4.3}$$

$$= \int_a^b f(r(t))\|r'(t)\|dt. \tag{4.4}$$

Example-(4.1):
Compute the line integral $\oint_c fds$ over the interval $0 \le t \le 2\pi$, where $f(x,y,z) = xz+y$ and the curve c is the helix parameterized by: $r(t) = (sint, cost)$.

Solution:
from the given parametric equation we can compute the velocity, and the speed as shown:

$$\begin{aligned} r(t) &= (sint, cost, t), then, \\ r'(t) &= (cost, -sint, 1) = velocity. \\ Speed &= \|r'(t)\| = \sqrt{cos^t + sin^2t + 1} = \sqrt{2}. \end{aligned}$$

where the components are:

$$\begin{aligned} x &= sint. \\ y &= cost. \\ z &= t. \end{aligned}$$

Then applying the line integral equation:

$$\oint_c fds = \int_a^b f(r(t))(\frac{ds}{dt})dt. \tag{4.5}$$

$$= \int_a^b f(r(t))\|r'(t)\|dt. \tag{4.6}$$

we find:

$$
\begin{aligned}
Interval &= [a, b] = [0, 2\pi], and \\
f(x, y, z) &= f(r(t)) = (sint, cost, z), then, \\
f(x, y, z) &= xz + y. \\
&= tsint + cost.
\end{aligned}
$$

Then equations (4.5), and (4.6) gives:

$$
\begin{aligned}
\oint fds &= \int_0^{2\pi} (tsint + cost)\sqrt{2} \; dt. \\
&= \sqrt{2} \; \{\int_0^{2\pi} t \; sint \; dt + \int_0^{2\pi} cost \; dt\}.
\end{aligned}
$$

In a similar way we can find the line integral of a vector field F as:

$$
\oint_c F.ds = \int_a^b F(r(t)).r'(t)dt. \tag{4.7}
$$

As an application for the line integral, we will see how it is used to find the work done by a force field in moving a particle along a curve. Where,

$$
Work = \int_c Fdr = \int_a^b F(x(t), y(t), z(t)).\frac{dr}{dt}dt. \tag{4.8}
$$

Example-(4.2):

Find the work done on a particle by a force moving upward along a circular helix parameterized by $: r(t) = (costi + sintj + tk)$ for $0 \leq t \leq 2\pi$, under a force $: F(x, y, z) = -xi - yj + 1/4k$.

Solution:

The parametric equations for the problem are:

$$
x(t) = cost.
$$

$$y(t) = sint.$$
$$z(t) = t.$$
$$And \ the \ force \ is: \ F(x,y,z) = -xi - yj + 1/4k.$$
$$= -costi - sintj + 1/4k.$$
$$= F(t).$$
$$r(t) = (cost, sint, t).$$
$$r'(t) = (-sint, cost, 1).$$
$$= \frac{dr}{dt}.$$
$$\Rightarrow dr = (-sint, cost, t)dt.$$

Computing the work, we get:

$$W = \int_C F.dr = \int_a^b F(x(t), y(t), z(t)).r'(t)) \ dt, \ where \ dr = r'dt.$$
$$= \int_0^{2\pi} (-costi - sint \ j + 1/4k)$$
$$. \ (-sinti + costj + tk) \ dt.$$
$$= \int_0^{2\pi} (cost \ sint - cost \ sint + 1/4)dt.$$
$$= 1/4 \int_0^{2\pi} dt = \frac{\pi}{2}.$$

Example-(4.3):
Compute $\oint_c xdx - ydy$, on the curve c parameterized by:
$r(t) = (cost, sint),$ where $0 \le t \le 2\pi$.

Solution:
From the given parametric equation we can find the followings:

$$r(t) = (cost, sint).$$
$$r'(t) = (-sint, cost).$$
$$Then: x = cost \Rightarrow x'(t) = dx = -sint.$$

$$And, y = sint \Rightarrow y'(t) = dy = cost.$$
$$Then, F(x(t), y(t)) = costi + sintj.$$

Then the integral becomes;

$$\oint F.ds = \int_a^b F(x(t), y(t))\ r'(t)\ dt.$$
$$= \int_0^{2\pi} (cost(-sint) + sint)dt.$$

Example-(4.4):

Compute: $\int_C x^2 y ds$, where C is the smooth closed curve shown in the fig(40).

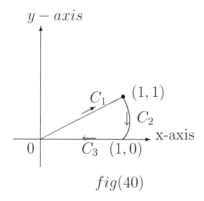

$$fig(40)$$

Solution:

The graph has 3-different paths, to compute the given integral over the curve means we have to compute the line integral on each path separately as follows: First we will describe and parametrize each curve, by letting $x = t$:

1.

$$C_1 \Rightarrow \begin{cases} x = t & 0 \le t \le 1 \\ y = t \end{cases}$$

Using the formula for ds we can compute the ds for C_1:

$$ds = \sqrt{(x'(t))^2 + (y'(t))^2} \, dt.$$
$$= \sqrt{1+1} \, dt = \sqrt{2} \, dt.$$

Thus, the given integral over C_1 is:

$$\int_{C_1} x^2 y \, ds = \int_0^1 t^3(\sqrt{2}) dt = \frac{\sqrt{2}}{4}. \tag{4.9}$$

2.

$$C_2 \Rightarrow \begin{cases} x = cost & \pi/4 \le t \le 0 \\ y = sint \end{cases}$$

Using the formula for ds we can compute the ds for C_1:

$$ds = \sqrt{(x'(t))^2 + (y'(t))^2} \, dt.$$
$$= \sqrt{(-sivt)^2 + (cost)^2} \, dt = dt.$$

The given integral over C_2 is:

$$\int_{C_2} x^2 y \, ds = \int_{\pi/4}^0 \cos^2 t \, \sin t \, dt. \tag{4.10}$$

To solve (4.10) we can use the substitution method as follows:
Let $u = cost \Rightarrow du = -sintdt.$, and the intervals changed to;
when $t = 0 \Rightarrow u = cos0 = 1$, and when $t = \pi/4 \Rightarrow u = cos45 = 1/\sqrt{2}$
Now the integral becomes:

$$\int_{C_2} x^2 y \, ds = \int_{\pi/4}^0 \cos^2 t \, \sin t \, dt.$$
$$= \int_1^{1/\sqrt{2}} u^2(-du).$$
$$= -\frac{u^3}{3}\Big|_1^{1/\sqrt{2}}.$$
$$= \frac{3}{2\sqrt{3}} - \frac{1}{3}.$$

3.

$$C_3 \Rightarrow \begin{cases} x = 1 - t & 0 \le t \le 1 \\ y = 0 \end{cases}$$

Using the formula for ds we can compute the ds for C_1:

$$\begin{aligned} ds &= \sqrt{(x'(t))^2 + (y'(t))^2} \, dt. \\ &= \sqrt{1 + 0} \, dt = dt. \end{aligned}$$

Thus, the given integral over C_3 is:

$$\int_{C_1} x^2 y \, ds = \int_0^1 t(0) dt = 0. \tag{4.11}$$

Now to compute the given integral we have to take the sum of all three curves;

$$\begin{aligned} \int_C x^2 y \, ds &= \int_{C_1} + \int_{C_2} + \int_{C_3}. \\ &= \frac{\sqrt{2}}{4} + \left\{ \frac{3}{2\sqrt{3}} - \frac{1}{3} \right\} + 0. \\ &= \frac{\sqrt{2}}{4} - \frac{3}{2\sqrt{3}} + \frac{1}{3}. \end{aligned}$$

Example-(4.5):
Compute the line integral $\oint_c f \, ds$ over the interval $0 \le t \le 2\pi$ where $f(x, y, z) = xz + y$ and the curve c is helix parameterized by: $r(t) = (sint, cost, t)$.
Solution:
Since we have the curve parameterized, then we can write the integral as:

$$\oint_c f \, ds = \int_a^b f(r(t)) \|r'(t)\| dt. \tag{4.12}$$

Now we will find the following:

$$Interval = [0, 2\pi].$$

$$\begin{aligned} f(x,y,z) &= xz+y. \\ f(r(t)) &= t\sin t + \cos t. \\ r(t) &= (\sin t, \cos t, t). \\ r'(t) &= (\cos t, -\sin t, 1). \\ \|r'(t)\| &= \sqrt{2}. \end{aligned}$$

Then (4.12) becomes:

$$\oint_c f ds = \int_0^{2\pi} (t\sin t + \cos t)\sqrt{2}dt = -2\pi\sqrt{2}. \qquad (4.13)$$

4.2   Surface Integrals

In section(4.1) we introduced the line integral of a function (integral over a curve) as a generalization of the ordinary single integral. In this section, we introduce the surface integral of a function (integral over a surface) as a generalization of the double integral. The surface integral of F over S defined as:

$$\int\int_S F(x,y,z)dA = \int\int_R F(x,y,f(x,y))\sqrt{f_x^2 + f_y^2 + 1}\ dA. \qquad (4.14)$$

Where,

$$\begin{aligned} S &= graph\ of\ f. \\ f &= defined\ function\ on\ the\ region R. \\ R &= region\ of\ f. \\ F &= continuous\ function\ defined\ on\ S. \\ f(x,y) &= z. \\ n &= <f_x, f_y, -1>, and \\ \|n\| &= \sqrt{f_x^2 + f_y^2 + 1}, with\ |n.k| = 1. \end{aligned}$$

To convert the surface integral into a double integral we will substitute : $z = f(x,y)$ in the function $F(x,y,z)$,and replace

the surface area element,
$dS \rightarrow |n|dA$, which for the surface $z = f(x,y)$ is given by :

$$dS = |n|dA = \sqrt{f_x^2 + f_y^2 + 1}dA.$$

For example: For the plane: $2x + y + z = 2$, $n = <2,1,1>$, and $|n| = \sqrt{6}$.

In this section we turn to a higher generalization similar to that of line integral, called surface integral. Surface integral help to describe certain physical phenomena and appear in Stokes and Divergence theorem from fundamental theorem of Calculus.

Example-(4.6):
Find $\int\int_S xyz\, dA$ where, S is part of the plane $x + y + z = 2$ in the first octant $(x \geq 0, y \geq 0, z \geq 0)$.

Solution:
To apply (4.14), first we need to determine the following partial derivatives first:

$$\begin{aligned} f(x,y) &= z = 2 - x - y. \\ f_x &= -1 \Rightarrow f_x^2 = 1. \\ f_y &= -1 \Rightarrow f_y^2 = 1. \\ \text{And, } \sqrt{f_x^2 + f_y^2 + 1} &= \sqrt{3}. \end{aligned}$$

$$z = 2 - x - y$$

$$fig(41)$$

Then equation (4.14) gives:

$$\int\int_S xyz\ dA = \int\int xy(2-x-y)\sqrt{3}\ dA. \qquad (4.15)$$

Then to find the intervals: I_x: let y=z=0, then $x = 2$. Then $I_y = f(x)$ let $z = 0$ and solve for y, gives: $y = 2 - x$.
Then the intervals are:

$$I_x\ :\ [0,2].$$
$$I_y\ :\ [0,(2-x)].$$

$$Then,\ \int\int_S xyz\ dA = \sqrt{3}\int_0^2\int_0^{2-x} xy(2-x-y)\ dy\ dx.$$

$$\int\int_S xyz\ dA = \frac{52\sqrt{3}}{45} \approx 2.$$

> **Homework**: Prove the result of equation (4.15).

Example-(4.7):

Evaluate the integral $\int\int_S f dA$, Where the surface S that is part of the plane $2x + 3y + 6z = 12$, located in the first octant.

Solution:
The equation of the plane, gives:

$$z = f(x,y) = \frac{12 - 2x - 3y}{4} = 3 - 1/2x - 3/4y.$$
$$f_x = -1/3;\ f_y = -1/2.$$
$$\sqrt{f_x^2 + f_y^2 + 1} = \frac{7}{6}.$$

Then the surface integral gives:

$$\int\int_S (6-2x)\ dA = \int\int_R \frac{36-12x}{7}\cdot\frac{7}{6}\ dA. \qquad (4.16)$$

$$= \int_0^6 \int_0^{\frac{12-2x}{3}} dy\ dx. \tag{4.17}$$

$$\int\int_S (6-2x)\ dA\ =\ 24. \tag{4.18}$$

4.3   Exercise - 1

1. Evaluate the surface integral S for $F(x, y, z) = xz + xy$ and S is the surface of cylinder $x^2 + y^2 = 16$ located in the first octant between $z = 0$, and $z = 4$.

2. Find the surface integral $\int \int_S (x + yz)\ dA$, where S is the portion of the plane $x + 2y + z = 2$ in the first octant.

3. Evaluate the surface integral for $F(x, y, z) = xz$, and S is in the first octant of the cylinder $x^2 + y^2 = 4$ between $z = 0$, and $z = 3$.

4. Evaluate the surface integral for $z = 2 - x^2 - y^2$, and S is in the first octant of the paraboloid $z = 2 - x^2 - y^2$ that is inside the cylinder $x^2 + y^2 = 4$.
See solutions on page-191.

4.4   Green's Theorem

Historic Notations

Green's theorem was named after George Green(1793-1841) the English Mathematician who published the theorem in 1828 , at age 35, in an essay on the application of Mathematical Analysis to the theories of Electricity and Magnetism, he was the only one who could explain the electrical phenomena at the time. He start his Mathematics Study at Cambridge College in 1838 at age 40. His theory is used in physics and Math in both pure and applied Math. Green's Theorem provides a relationship between a Line integral around a simple closed curve ($\int_C$) and a double integral over the region ($\int \int_R$) enclosed by the curve. The direction of the curve is determined by the parameterization, and the positive parameter value that is in the counterclockwise direction as shown in fig(42).

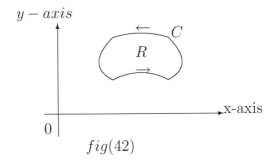

$$fig(42)$$

Greens Theorem: Let the simple closed curve C be oriented in the counter clockwise direction, and let R be the region enclosed by C. If f, and g have continuous first partial derivative in an open set containing R, then:

$$\int f(x,y)dx + g(x,y)dy = \int\int_R (\frac{\partial g(x,y)}{\partial x} - \frac{\partial f(x,y)}{\partial y})dA.$$

(4.19)

If we let: $f(x,y) = M$, and $g(x,y) = N$, then equation (4.19) can be written in terms of M, and N as follows:

$$\int_C Mdx + Ndy = \int\int_R (\frac{\partial N}{\partial x} - \frac{\partial M}{\partial y})dA.$$

(4.20)

Equations (4.19), and (4.20) are called Green's Theorem equations.

Homework: Prove equation (4.20).

4.5 Greens Theorem in Vector Form

We have used Greens theorem as:

$$\int_c fdx + gdy = \int\int_R (g_x - f_y)dA.$$

(4.21)

Using curl, and Divergent operators, we see the left side of equation (4.21) is:

$$\oint_c fdx + gdy = \oint_c F.dr,$$

(4.22)

and from the right side of equation(4.23) we see:

$$(\frac{\partial g}{\partial x} - \frac{\partial f}{\partial y})k = curlF = \nabla \times F = \begin{vmatrix} i & j & k \\ \frac{\partial}{\partial x} & \frac{\partial}{\partial y} & \frac{\partial}{\partial z} \\ f & g & 0 \end{vmatrix}$$

(4.23)

Then based on the curl and divergent operators, Greens theorem can be written as:

$$\oint_c F.dr = \int\int_R (\nabla \times F).k dA. \qquad (4.24)$$

Equation (4.24) is called Green's Theorem in Vector Form.

Example-(4.8): Evaluating Linear integral using Green's theorem:

Use Green's Theorem to evaluate the Line Integral:
$\int_C (e^{-x^2} + x^2 y)dx + (x^2 + \sqrt{1+y^3})dy$. Where, C is the path in the counterclockwise direction as given in the fig(42).

Solution:
From the given line integral we get the following functions:

$$f(x,y) = e^{-x^2} + x^2 y \Rightarrow \frac{\partial f}{\partial y} = x^2.$$

$$g(x,y) = x^2 + \sqrt{1+y^3} \Rightarrow \frac{\partial g}{\partial x} = 2x.$$

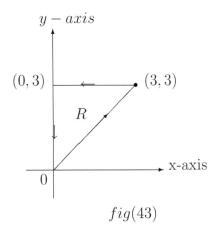

$$fig(43)$$

Now applying Greens Formula we can write :

$$\int_C f\,dx + g\,dy = \int\int_R (g_x - f_y)\,dA.$$

$$\int_C (e^{-x^2} + x^2 y)\,dx + (x^2 + \sqrt{1 + y^3})\,dy = \int\int_R (2x - x^2)\,da.$$

$$= \int_0^3 \int_0^y (2x - x^2)\,dx\,dy.$$

$$= \cdots \int_0^y (2x - x^2)\,dx \cdots.$$

$$= \int_0^3 (y^2 - \frac{y^3}{3})\,dy.$$

$$\text{Then, } \int_C f\,dx + g\,dy = \frac{27}{12} = 2.25.$$

Note: The above curve is a simple one and of both Type-I, and Type-II, if the strip method is used taking both horizontal strip with thickness dy or vertical strip with thickness dx will solve the problem.

Example-(4.9):

Use Green's Theorem to evaluate the Line Integral on C :
$\int_C (x^2 y - 2y^2)\,dx + (x^3 - 2y^2)\,dy$, oriented in the counterclockwise direction, where C is the rectangle with the vertices as given in the fig below:

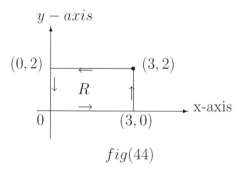

$$fig(44)$$

Solution:

First we compute the following partial derivatives:

$$f(x,y) \;=\; x^2y - 2y^2 \rightarrow f_y = x^2 - 4y.$$
$$g(x,y) \;=\; x^3 - 2y^2 \rightarrow g_x = 3x^2.$$

Then applying Greens Theorem gives:

$$
\int_C (x^2y - 2y^2)dx + (x^3 - 2y^2)dy \;=\; \int_0^2 \int_0^3 (g_x - f_y)dxdy.
$$
$$
= \int_0^2 \int_0^2 (3x^2 - (x^2 - 4y))dxdy.
$$
$$
= \int_0^2 \int -0^3(2x^2 + 4y)dxdy.
$$
$$
= \int_0^2 (18 + 12y)dy.
$$
$$
= 12.
$$

4.6   Exercise - 2

1. Use Greens theorem to evaluate $\int_c F.dr$, where C is a circle : $x^2 + y^2 = 4$, oriented counterclockwise, and $F = yi + 2xyj$.

2. Use Greens Theorem to evaluate the integral: $\oint_c y^2 dx + x^3 dy$, where c is the path formed by the square with vertices on ; $(0,0), (2,0), (2,2), (0,2)$, on clockwise direction.

3. Use Greens Theorem to evaluate $\int_c F.dr$, where C is counterclockwise, $F(x,y) = yi + 3xj$. See solutions on page - 191.

4.7 Divergence Theorem

The Divergence Theorem, also known as Gauss Theorem, relates the integral of div.F over a 3-dimensional region (R^3) to the outward flux of F over the closed boundary surface S.

$$\int\int_S F.nds = \int\int\int_R (\nabla \times F).kdA. \qquad (4.25)$$

In the previous section we found Greens theorem in vector form as:

$$\oint_c F.Nds = \int\int_R \nabla.F(x,y)dA. \qquad (4.26)$$

Where, $F \in R^2$. We can extend this theorem taking $F \in R^3$ as:

$$\oint_c F.nds = \int\int\int_R \nabla.F(x,y,z)dV. \qquad (4.27)$$

Taking the normal vector n to be outward unit normal to the solid,

$$\oint_c F.nds = \oint F.ds. \qquad (4.28)$$

Then the divergence theorem becomes:

$$\int\int_S F.ds = \int\int\int_R \nabla.F(x,y,z)dV. \qquad (4.29)$$

Homework: Prove equation (4.29).

Example-(4.10):

Evaluate $\int\int_S F.ds$, where $F(x, y, z) = 2xi + xyj + 3xzk$;
R is in the rectangular box with the following dimensions:
$x = 0, x = 1, y = 0, y = 1, z = 0, z = 2$.

Solution:

$$
\begin{aligned}
\nabla.F &= (\frac{\partial}{\partial x}i + \frac{\partial}{\partial y}j + \frac{\partial}{\partial z}k).(2xi + xyj + 3xzk). \\
&= \frac{\partial}{\partial x}(2x) + \frac{\partial}{\partial y}(xy) + \frac{\partial}{\partial z}(3xz). \\
&= 2 + x + 3x = 2(1 + 2x).
\end{aligned}
$$

$R = \{(x, y, z)|0 \le x \le 1, 0 \le y \le 1, 0 \le z \le 2\}$. Then,

$$
\begin{aligned}
\int\int_s F.ds &= \int\int\int_R 2(1 + 2x)dV. \\
&= 2\int_0^2 \int_0^1 \int_0^1 (1 + 2x)dxdydz. \\
&= 8.
\end{aligned}
$$

4.8   Exercise - 3

1. Use the divergence theorem to evaluate $\int\int_S F.nds$ where
$F(x, y, z) = xi + yj - z^2k$, and S is the sphere : $x^2 + y^2 + z^2 = 4$.

2. Use the divergence theorem to evaluate $\int\int_S F.nds$ where
$F(x, y, z) = 2x^2i - y^2xj + x^2zk$, and S is the surface of the
rectangular box bounded by the planes: $0 \le x \le 2, 0 \le y \le 1$,
$0 \le z \le 3$.

3. Compute $\int\int_S F.ds$ where $F(x, y, z) = -2xi + 3yj - 4zk$,

bounded by the plane: $x + 2y + 2z = 4$.

4. Compute $\int \int_S F.ds$ where $F(x, y, z) = y^2 zi + zxj + x^2 k$; S is the boundary of the solid region bounded by the planes: x=1, and y+z $= 1$, and the coordinate planes.

See solutions on page - 192.

4.9 Stokes' Theorem

We have seen how Greens theorem relates the line integral of the curve bounding the region R to the double integral over the region R, given as:

$$\int_C f(x,y)dx + g(x,y)dy = \int \int_R (\frac{\partial g(x,y)}{\partial x} - \frac{\partial f(x,y)}{\partial y})dA.$$

$$(4.30)$$

Now, we'll see Stokes Theorem relating the line integral around the boundary curve S to the double integral over the surface S, given as:

$$\int_C F.dr = \int \int_S (\nabla \times F).dS. \qquad (4.31)$$

Where S is a smooth surface, and the unit vector n is normal to S and pointing "UP" or toward positive z.

We may write the normal n as $n = (cos\alpha, cos\beta, cos\gamma)$, where $,\alpha, \beta,$ andγ are angles from the normal to the positive x, y axis as shown in $fig(45)$. Since normal points up then $cos\gamma \geq 0$ (acute angle).

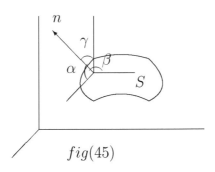

$$fig(45)$$

Example-(4.11):
Use Stokes theorem to evaluate $\oint_C F.dr$ for: $F(x, y, z) = (4x - 1/2y)i - 3yz^2j - 3y^2zk$, S is the upper surface of the sphere $x^2 + y^2 + z^2 = 1$, and C is the boundary.

Solution:
Using Stokes Equation given in (4.33), we have to find the following first:

$$\nabla \times F = \begin{vmatrix} i & j & k \\ \frac{\partial}{\partial x} & \frac{\partial}{\partial y} & \frac{\partial}{\partial z} \\ (4x - 1/2y) & -3yz^2 & -3y^2z \end{vmatrix} = 1/2k. \quad (4.32)$$

Now evaluating the surface integral for the problem we get:

$$\int\int_S (\nabla \times F).n \ ds \ = \ \int\int_S 1/2k.ndA. \quad (4.33)$$

$$= \ 1/2 \int\int_S dA. \quad (4.34)$$

$$= 1/2 \int_{-2}^{2} \int_{-\sqrt{4-x^2}}^{\sqrt{4-x^2}} \sqrt{4-x^2} dx. (4.35)$$

$$= 2\pi. \tag{4.36}$$

Evaluating the line integral we need the parametric components:

$$x = 2cost.$$

$$y = 2sint.$$

$$z = 0, taking \ the \ circle \ in \ xy - plane.Then:$$

$$\oint_C F.dr = \oint_C (4x - 1/2y)dx - 3yz^2dy - y^2zdz. \tag{4.37}$$

$$= \int_0^{2\pi} (4cost - 1/2sint)(-sint)dt. \tag{4.38}$$

$$= 2\pi. \tag{4.39}$$

We see that the integrals: $\oint_C$ of (4.39) $= \int \int_S$ of (4.36) which proves Stokes theorem.

4.10   Gradient, Divergence, Curl and Laplacian

Gradient:

The gradient operator or "del operator" is denoted by " $\bigtriangledown$ ", it works as a differential operator and as a vector. In Cartesian Coordinates on R^3 is defined as:

$$\bigtriangledown = i\frac{\partial}{\partial x} + j\frac{\partial}{\partial y} + k\frac{\partial}{\partial z}. \tag{4.40}$$

If applied to a function f in R^3, at a point $p(x, y, z)$ is denoted by $\bigtriangledown f(x, y, z)$, and it is read as (grad-f). The expansion formula is:

$$\nabla f(x, y, z) = (i\frac{\partial}{\partial x} + j\frac{\partial}{\partial y} + k\frac{\partial}{\partial z})f(x, y, z).$$
$$= i\frac{\partial f}{\partial x} + j\frac{\partial f}{\partial y} + k\frac{\partial f}{\partial z}.$$
$$= if_x + jf_y + kf_z.$$

Provided that the partial derivatives exists. The del-operator can also be defined in R^n for any n-coordinates given as,

$$\nabla = e_1\frac{\partial}{\partial x_1} + e_2\frac{\partial}{\partial x_2} + e_3\frac{\partial}{\partial x_3} \cdots e_n\frac{\partial}{\partial x_n}. \tag{4.41}$$

Where, $e_i = (0 \cdots 1, \cdots 0)$, $i = 1, \cdots, n$ is the standard basis vector in R^n.

Example-(4.12):
Find ∇f at the point $(3, 2)$ for $f(x, y) = ln(xy)$.

Solution:

$$f(x, y) = ln(xy) \Rightarrow lnx + lny.$$
$$f_x = 1/x.$$
$$f_y = 1/y.$$
$$\nabla f = (f_x, f_y) = (1/x, 1/y).$$
$$Then, \nabla f(3, 2) = (1/3, 1/2).$$
$$= 1/3j + 1/2j.$$

Divergence:

Applying the "del-operator" directly to a scaler field f is called " gradient of scalar" denoted by ∇f and results into a vector field, as shown: Let $f = 3x + 2y^2 - 4z^3$, which is a scalar. Applying the del-operator gives:

$$\nabla f = i\frac{\partial}{\partial x}(3x) + j\frac{\partial}{\partial y}(2y^2) + k\frac{\partial}{\partial z}(-4z^3).$$

$$\nabla f = 3i + 4yj - 12z^2k = Vector.$$

But applying the "del-operator" to a vector field F is called " div F" denoted by $\nabla.F$, and results into a scalar field. Let $F = 3xi + 2y^2j - 4z^3k$, which is a scalar. Applying the del-operator gives:

$$\nabla F = (i\frac{\partial}{\partial x} + j\frac{\partial}{\partial y} + k\frac{\partial}{\partial z}).(3xi + 2y^2j - 4z^3k).$$

$$= \frac{\partial}{\partial x}(3x) + j\frac{\partial}{\partial y}(2y^2) + k\frac{\partial}{\partial z}(-4z^3).$$

$$= 3x + 4y - 12z^2 = scalar.$$

Curl of a vector field:

Let $F(x, y, z) = fi + gj + hk$ then "curl-F" denoted by $\nabla \times F$, results into a vector field as shown:

$$curlF = \nabla \times F = \begin{vmatrix} i & j & k \\ \frac{\partial}{\partial x} & \frac{\partial}{\partial y} & \frac{\partial}{\partial z} \\ f & g & h \end{vmatrix}.$$

$$= i(\frac{\partial h}{\partial y} - \frac{\partial g}{\partial z}) - j(\frac{\partial h}{\partial x} - \frac{\partial f}{\partial z}) + k(\frac{\partial g}{\partial x} - \frac{\partial f}{\partial y}).$$

$$= i(h_y - g_z) - j(f_z - h_x) + k(g_x - f_y)$$

$$= Vector.$$

Laplacian:

If f is a scalar whose second partial derivative exists, the Laplacian of f denoted by " $\nabla^2 f$" is defined as:

$$\nabla^2 f = \frac{\partial^2 f}{\partial x^2} + \frac{\partial^2 f}{\partial y^2} + \frac{\partial f}{\partial z^2}. \tag{4.42}$$

$$= f_{xx} + f_{yy} + f_{zz}. \tag{4.43}$$

The same result can be accomplished by taking the gradient of del-f,

$$\nabla^2 f = div(grad - f) = \nabla.(\nabla f). \tag{4.44}$$

$$= (\frac{\partial}{\partial x}i + \frac{\partial}{\partial y}j + \frac{\partial}{\partial z}k).(f_x i + f_y j + f_z k). \tag{4.45}$$

$$= \frac{\partial f_x}{\partial x} + \frac{\partial f_y}{\partial y} + \frac{\partial f_z}{\partial z}. \tag{4.46}$$

$$= f_{xx} + f_{yy} + f_{zz}. \tag{4.47}$$

When $\nabla^2 f = 0$, it is known as Laplacian equation which is Harmonic function that is used in physics.

Example-(4.13):

Find div F in r^3 if $F(x, y, z) = 2zi + xj + y^3 k$.

Solution: $div F = \nabla.F = \frac{\partial}{\partial x}(2z) + \frac{\partial}{\partial y}(x) + \frac{\partial}{\partial z}(y^3) = 0.$

Example-(4.14):
For the vector $F(x, y, z) = (2x-y)i-(2y-z)j+zk$. Find div F.

Solution: $div F = \nabla.F = \frac{\partial}{\partial x}(2x-y)+\frac{\partial}{\partial y}(-2y-z)+\frac{\partial f_z}{\partial z}(z) = 1.$

Example-(4.15):

Find curl F , if F is the vector field given as:
$$F(x, y, z) = (y - 2z)e^x i + e^x j - 2e^x k.$$

Solution:

$$curl F = \nabla \times F = \begin{vmatrix} i & j & k \\ \frac{\partial}{\partial x} & \frac{\partial}{\partial y} & \frac{\partial}{\partial z} \\ (y - 2z)e^x & e^x & -2e^x \end{vmatrix}.$$

$$= e^x(1 - y)k.$$

$$Then, \; curl F = vector \; on \; z - direction.$$

4.11   Some Theorems and Proofs

In this section we include some proofs that might be helpful for the student to go over, as a review for the material studied in Calculus-II, and assistance for Vector Calculus.

The Sandwich Theorem :

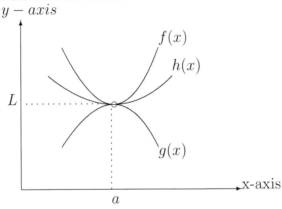

$$fig(46)$$

Suppose $h(x)$ is a function located as shown in fig(46), such as: $f(x) \leq h(x) \leq g(x)$, and for all $x \neq a$,

$$lin_{x \to a} f(x) = lim_{x \to a} = L, \tag{4.48}$$

then: $lim_{x \to a} h(x) = L.$

$$then : \quad lim_{x \to a} h(x) = L. \tag{4.49}$$

Proof

Assuming that the 2-sided limits of both functions $f(x)$, and $g(x)$ equal some constant L as x approaches a, this means, for some $\in > 0$ there exists $\delta > 0$ such that:

$$L- \in < f(x) \leq h(x) \leq g(x) < L+ \in .$$

$$Then, L- \in \quad < \quad h(x) < L+ \in .$$
$$Then \ for \ all \ x: \quad c < x < c + \delta \quad \Rightarrow \quad \|h(x) - L\| < \in .$$
$$then: \quad lim_{x \to c^+} h(x) = L.$$

In a similar way it can be shown that : $lim_{x \to c^-} h(x) = L$.
If the 2-sided limits for $h(x)$ approach L, then equation (4.49)
is satisfied.

The Zipper Theorem

If the sequence $\{a_n\}$, and $\{b_n\}$ both converge to L, then the
sequence, $a_1, b_1, a_2, b_2, \cdots, \cdots, a_n, b_n$, also converges to L.
The proof can be done in a similar way to the squeeze theorem.

The Divergence Theorem of Gauss

The Divergence theorem states that: The surface Integral of
the normal component of a vector taken over a closed surface
is equal to the integral of the divergence of the vector over the
volume enclosed by the surface:

$$\int \int_S F.ds = \int \int_S F.nds = \int \int \int_V \nabla.FdV. \qquad (4.50)$$

Where n is the positive (outward) normal to S. The divergence
Theorem is a generalization of Green's Theorem.

Stokes Theorem

Stokes Theorem states that: The line integral of the tangential
component of a vector taken around a simple closed curve is
equal to the surface integral of the normal component of the
curl of the vector taken over any surface having the curve as
its boundary.

$$\oint_C F.dr = \int\int_S (\nabla \times F).ndS = \int\int_S (\nabla \times F).dS. \quad (4.51)$$

Where C in counterclockwise direction.

Green's Theorem in the plane

Green's Theorem relates a closed region in R to a closed curve that bounds the region.

$$\oint_C Mdx + Ndy = \int\int_R \left(\frac{\partial N}{\partial x} - \frac{\partial M}{\partial y}\right)dxdy. \quad (4.52)$$

Green's Theorem in the plane is a special case of Stokes Theorem.

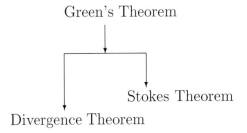

Green's Theorem

Stokes Theorem

Divergence Theorem

Notice how Greens Theorem is related to both Divergence Theorem and stokes theorem.

4.12   Exercise Solutions

Chapter-1

Exercise-1 on Page (21): Graphs of functions

1. The graph of the vector $u = i + 4j + 3k$ in R^3:

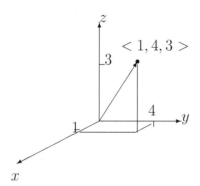

2. Graph of vectors in R^2:

a. $< -2, 2 >$

b. $< 3, 2 >$

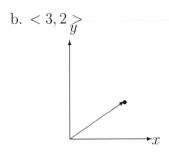

3. $u = 1/4 < 12, 4, 8 >=< 3, 1, 2 >$, and
$v = 2 < 1, 2, 3 >=< 2, 4, 6 >$. Then : $u + v =< 3, 1, 2 > + < 2, 4, 5 >=< 5, 5, 8 >$

6. $x + 6 = 4$ or $x = -2$, $8 + y = -2$, or $y = -10$. $5 - 8 = 7$, or $z = -3$.

7. Magnitude $=$ length $= \|u\| = \sqrt{5^2 + 3^2} = \sqrt{34}$'

8. Direction $=$ angle (θ) with the horizontal line.

$\theta = tan^{-1}(\frac{y}{x}) = tan^{-1}(\sqrt{3}) = 60°$.

9. Find the distance or Magnitude:
a. $u =< -4, 3 > \rightarrow \|u\| = \sqrt{(-4)^2 + 3^2} = 5$.
b. $u =< -22, 25 > \rightarrow \|u\| = \sqrt{(-22)^2 + 25^2} = 26.63$.
c. $u =< 8, -3 > \rightarrow \|u\| = \sqrt{(8)^2 + (-3)^2} = \sqrt{73}$.

10. unit vector $u^\wedge = \frac{u}{\|u\|}$. $u = -2i - 5j$, $v = 3i + 2j$.

a. $\|1/2u + v\| = \|1/2 < -2i - 5j > + < 3i + 2j > \|$.

$$= \| < -1, -5/2 > + < 3, 2 > \|.$$
$$= \| < 2, -0.5 > \|.$$
$$Then, \ \|1/2u + v\| = \sqrt{(2)^2 + (-0.5)^2} = \sqrt{17}/2.$$
$$And, \ the \ unit \ vector \ is = \frac{1/2u + v}{\|1/2u + v\|}.$$
$$= \frac{1}{\sqrt{17}} < 4, -1 > .$$

b.

$$\begin{aligned}
\|2u - v\| &= \|2 < -2, -5 > - < 3, 2 > \|. \\
&= \| < -4, -10 > - < 3, 2 > \|. \\
&= \| < -4 - 3, -10 - 2 > \|. \\
&= \| < -7, -12 > \| = \sqrt{193}.
\end{aligned}$$
$$\begin{aligned}
Then \ the \ unit \ vector \ is \ &= \ \frac{2u - v}{\|2u - v\|} \\
&= \ \frac{1}{\sqrt{193}} < -7, -12 > .
\end{aligned}$$

11.
a. $u = -2i - 3j.$

$$\begin{aligned}
\|u\| &= \sqrt{(-2)^2 + (-3)^2} = \sqrt{12}. \\
u^\wedge &= \frac{u}{\|u\|} = \frac{-2i + 3j}{\sqrt{13}} = \frac{1}{\sqrt{13}} < -2, 3 > .
\end{aligned}$$

b. $u = -5i + 9j.$

$$u^\wedge = \frac{u}{\|u\|} = \frac{-5i + 9j}{\sqrt{(-5)^2 + (9)^2}} = \frac{1}{\sqrt{106}} < -5, 9 > .$$

c. $u = i + 12j + 17k$.

$$u^\wedge = \frac{u}{\|u\|} = \frac{i + 12j + 17k}{\sqrt{(1)^2 + (12)^2 + (17)^2}} = \frac{1}{\sqrt{434}} < 1, 12, 17 > .$$

12. $F_1 = 300$ *pound.* $F_2 = 450$ *pound.*

$$\text{Then the resultant force } R = F_1 - F_2.$$
$$= 300 - 450$$
$$\text{Then, } R = -150 \text{ pound to the left.}$$

Exercise-2 on Page (34): Dot- Product

Compute: $u.v, \|u\|, and \|v\|$:
1. $u =< 2, 3 >, v =< -3, 5 >$:

$$u.v = < 2, 3 > . < -3, 5 >$$
$$= (2i + 3j).) - 3i + 5j).$$
$$= (2.3) + (-3.5).$$

Where, $i.i = 1,$ *and* $j.j = 1$.

$$= 6 - 15.$$
$$\text{Then } u.v = -9.$$

$$\|u\| = \sqrt{2^2 + 3^2}.$$
$$\text{And, } v = \sqrt{(-3)^2 + 5^2} = \sqrt{34}.$$

2. $u =< 2, -6, -4 >, v =< 1, -2, -5 >$:

$$u.v = 2(1) + (-6)(-2) + (-4)(-5).$$

$$= \ 2 + 12 + 20.$$
Then $u.v \ = \ 34.$

$$\|u\| \ = \ \sqrt{2^2 + (-6)^2 + (-4)^2} = \sqrt{56}.$$
And, $\|v\| \ = \ \sqrt{(1)^2 + (-2)^2 + (-5)^2} = \sqrt{30}.$

3.

$$u.v \ = \ -2(-1/2) + 3(3/2) + (-1)(1) = \frac{9}{2}.$$
$$\|u\| \ = \ \sqrt{14}.$$
$$\|v\| \ = \ \sqrt{3.5}.$$

4. $u.v = -3$, $\|u\| = 3$, $\|v\| = \sqrt{10}$.

5. Angle $\theta = cos^{-1}\left(\frac{u.v}{\|u\|\|v\|}\right) = 96.4°$.

6. Angle $\theta = cos^{-1}\left(\frac{u.v}{\|u\|\|v\|}\right) = 108.4°$.

7. Angle $\theta = cos^{-1}\left(\frac{u.v}{\|u\|\|v\|}\right) = 51.9°$.

8. $u^{\wedge} = \frac{u}{\|u\|} = < \frac{-2}{\sqrt{5}}, 0, \frac{1}{\sqrt{5}} >$.

11. Since $u.v = 0$, then $u, and v$ are perpendicular.

12. Since, $u.v = -2\sqrt{2} + 6 \neq 0$, then the two vectors are not perpendicular.

13. Since $u.v = u.\frac{u}{\|u\|} = 1$.

Exercise-3 on Page (46): Cross- Product

1. u x v = 2i - 10j + 12k .

2. u x v = 25.

3.

$$u \times v = \begin{bmatrix} i & j & k \\ 4 & 1 & 4 \\ -1 & 6 & 1 \end{bmatrix} = -23i - 8j + 25k$$

4. Let the vector w be orthogonal to both u, and v Then:

$$w = u \times v = \begin{bmatrix} i & j & k \\ 2 & 3 & -1 \\ 4 & 6 & 1 \end{bmatrix} = 9i - 6j$$

5. Let the vector w be orthogonal to both u, and v Then:

$$w = u \times v = \begin{bmatrix} i & j & k \\ 5 & 0 & -1 \\ 4 & 6 & 1 \end{bmatrix} = 18i - 34j + 30k$$

6. Let the vector w be orthogonal to both u, and v Then:

$$w = u \times v = \begin{bmatrix} i & j & k \\ 4 & -5 & 1 \\ 5 & 0 & 1 \end{bmatrix} = -5i + j + 25k$$

7.

$$u.(v \times w) = \begin{bmatrix} 4 & 1 & 3 \\ 3 & 4 & -1 \\ 5 & 4 & 1 \end{bmatrix} = 0$$

8.

$$u \times (v \times w) = < 4,1,3 > \times \begin{bmatrix} i & j & k \\ 3 & 4 & -1 \\ 5 & 4 & 1 \end{bmatrix}$$

$$= \begin{bmatrix} i & j & k \\ 4 & 1 & 3 \\ 8 & -8 & -8 \end{bmatrix}$$

$$= < 16,47,-40 > vector.$$

9.

$$(u \times v).(u \times w) = \begin{bmatrix} i & j & k \\ 4 & 1 & 3 \\ 3 & 4 & -1 \end{bmatrix} . \begin{bmatrix} i & j & k \\ 4 & 1 & 3 \\ 3 & 4 & -1 \end{bmatrix}$$

$$= < -13,13,11 > . < -11,11,11 >$$

$$= 407 = scalar.$$

10.

$$AB = B - A = < 4-5, 4-3, 1-2 > = < -1,1,-1 > .$$
$$AC = C - A = < 6-5, 2-3, 4-2 > = < 1,-1,2 > .$$

Then the volume is:

$$\|AB \times AC\| = \begin{bmatrix} i & j & k \\ -1 & 1 & -1 \\ 1 & -1 & 2 \end{bmatrix} = \sqrt{2}$$

11. Area of triangle $= 1/2 \|u \times v\|$. Then:

$$A = 1/2\|u \times v\| = 1/2 \begin{bmatrix} i & j & k \\ 2 & 3 & -4 \\ 1 & 4 & 0 \end{bmatrix}$$

$$= 1/2 < 16, -4, 5 >$$

$$= 8.62$$

12. Volume $= u.(v \times w)$. where:

$$
\begin{aligned}
AB &= <-2, 2, 3> = u. \\
AC &= <-5, 1, 5> = v. \\
AD &= <1, 1, -3> = w.
\end{aligned}
$$

$$
Then \ the \ Volume = u.(v \times w) = \begin{bmatrix} i & j & k \\ 2 & 3 & -4 \\ 1 & 4 & 0 \end{bmatrix}
$$

$$
= 22
$$

13. Vectors $u, v, and w$ are coplanar iff : $u.(v \times w) = 0$

$$
u.(v \times w) = \begin{bmatrix} 4 & -6 & 8 \\ 2 & 4 & -2 \\ 14 & 0 & 10 \end{bmatrix}
$$

$$
= 0
$$

14.

$$
\begin{aligned}
ab &= b - a = <2, -3, -1> . \\
ac &= c - a = -1, 2, 0. \\
ad &= 0, 1, -1. Then,
\end{aligned}
$$

$$
ab.(ac \times ad) = \begin{bmatrix} 2 & -3 & -1 \\ -1 & 2 & 0 \\ 0 & 1 & -1 \end{bmatrix}
$$

$$
= 0
$$

15. Let the vector be w, then:

$$
w = u \times v = \begin{bmatrix} i & j & k \\ 0 & 2 & -6 \\ 4 & 8 & -2 \end{bmatrix}
$$

$$= 44i - 24j - 8k.$$

Where, $u.w = v.w = 0$.

16. The angle between the two vectors is :

$$\theta = \cos^{-1}\left(\frac{u.v}{\|u\|\|v\|}\right).$$
$$= \cos^{-1}\left(\frac{<0,2,-6>.<4,8,-1>}{\sqrt{4+36}\sqrt{16+64+4}}\right).$$
$$Then, \theta = \cos^{-1}\left(\frac{28}{\sqrt{3360}}\right) = 61°$$

17. Projection of u onto v is:

$$Proj_v u = \left(\frac{v.u}{\|v\|^2}\right)u.$$
$$= \frac{-13}{21}<-3,-1,-1>.$$

Exercise-4 on Page (58): Lines and Planes.

1. Given: $p = (-3, 2, 1)$, and $u = -3i + j + 2k$.
Then using the parametric equation gives:

$$r = r_0 + tu.$$
$$r = (-3, 2, 1) + t<-3, 1, 2>.$$
$$Then, r = <-3 - 3t, 2 + t, 1 + 2t>.$$

And its symmetric equations are:

$$x = -3 - 3t \Rightarrow \frac{x+3}{-3}.$$

$$y = 2 + t \Rightarrow \frac{y-2}{1}.$$

$$z = 1 + 2t \Rightarrow \frac{z-1}{2}.$$

And, the symmetric equation is:

$$\frac{x+3}{-3} = \frac{y-2}{1} = \frac{z-1}{2}.$$

2.

$$r = r_0 + tu.$$
$$= (3,0,6) + t(3j + 4k).$$
$$r = \ <3, 3t, 6 + 4t>.$$

And the symmetric equation is:

$$3 = \frac{y}{3} = \frac{z-6}{1}.$$

3. $r = <-3 - 3t, 2 + t, 1>$, and the symmetric equation is:

$$\frac{x+3}{-3} = \frac{y-2}{1} = \frac{z-1}{1}.$$

4. From the given equation of the parallel line:
$a = 5, b = 5, c = 1$, then:

$$x = 3 + 5t.$$
$$y = -2 + 5t.$$
$$z = 2 + t.$$

5. Let $L_1 = u = p_1 p_2 = p_2 - p_1 = (-3 - 4, -5 - 3, 7 - 4)$
$=< -7, -8, 3 >.$
And , $L_2 = v = p_1 p_2 = p_2 - p_1 = (-4 + 1, 5 - 3, 2 - 3)$
$=< -3, 2, -1 >.$

Since $u \neq cv$, then the lines are not parallel.
The lines are perpendicular iff: $u.v = 0$.

6. From the symmetric equation we get: $\frac{x-1}{-3} = \frac{y-2}{12}, z = 5$.
Then: $L_1 = u =< -3, 12, 1 >$.
And, $L_2 = v = p_1 p_2 = p_2 - p_1 =< 4 - 5, 11 - 7, 9 - 9 >$
$=< -1, 4, 0 >$.
Since $u =< -3, 12, 0 >= 3 < -1, 4, 0 >$. Or $u = 3v$.
Then u, *and* v are skewed vectors.

Exercise-5 on Page (73): Curvilinear-Coordinates.
1. Find the Cartesian coordinates for the given polar coordinates: $(\sqrt{2}, \frac{\pi}{3})$.
From the given polar coordinates $(\sqrt{2}, \frac{\pi}{3})$: $r = -\sqrt{2}$,
and $\theta = \pi/3$. Then,

$$x = r\cos\theta = \sqrt{2}\cos\frac{\pi}{3} = \sqrt{2}\frac{\sqrt{3}}{2}.$$

$$y = r\sin\theta = \sqrt{2}\cos\frac{\pi}{3} = \frac{\sqrt{6}}{2}.$$

$$Then\ (r, \theta) = (\sqrt{2}, \pi/3) \Rightarrow (x, y) = (1/2, \sqrt{3}/2).$$

2. For $(x, y) = (3, 2)$ find (r, θ).

Given is: $x = 3 = r\cos\theta$, and $y = 2 = r\sin\theta$.

Using Pythagorean theorem gives: $r = \sqrt{13}$.

And $\theta = tan^{-1}(\frac{y}{x}) = tan^{-1}(\frac{3}{2}) = 56.3°$.

Then : $(x, y) = (3, 2) \Rightarrow (\sqrt{13}\cos 56.3, \sqrt{13}\sin 56.3)$.

3. Change $(r, \theta, z) = (1/.2, \pi/3, 2)$ into (x, y, z).

From the given cylindrical coordinates: $(r, \theta, z) = (1/2, \pi/3, 2)$:

$r = 1/2, \theta = \pi/2, z = 2$. Then ,

$x = r\cos\theta = 1/2\cos 60 = 1/4$.

And $y = r\sin\theta = 1/2\sin 60 = \sqrt{3}4$.

Then $(r, \theta, z) = (1/2, \pi/3, 2) \Rightarrow (x, y, z) = (1/4, \sqrt{3}/4, 2)$.

4. Convert from spherical coordinates $(\rho, \theta, \phi) = (4, \pi/2, \pi/4)$ to Cartesian.

From the given spherical coordinates: $\rho = 4, \theta = \pi/2 = 90°$, and
$\phi = \pi/4 = 45°$. Then,

$x = \rho \sin \phi \cos \theta = 4 \sin 45 \cos 90 = 0$.

$y = \rho \sin \phi \sin\theta = 4 \sin 45 \sin 90 = 2\sqrt{2}$.

$z = \rho \cos \phi = 4 \cos 45 = 2\sqrt{2}$.

Then $(r, \theta, z) = (1/2, \pi/3, 2) \Rightarrow (x, y, z) = (0, 2\sqrt{2}, 2\sqrt{2})$.

5. From the given Cartesian $= (2, \sqrt{3}, 3) = (x, y, z)$. we get,
$x = 2, y = \sqrt{3}, z = 3$, then $r = \sqrt{13}$,
and $\theta = tan^{-1}(y/x) = tan^{-1}(\sqrt{3}/2) = 40.9^{circ}$. And $z = 3$.
Then,
$(r, \theta, z) = \sqrt{13}, 40.9°, 3)$.

Exercise-6 on Page (77): Vector-Valued Functions.

1.

$$F(t) = \begin{bmatrix} i & j & k \\ t-1 & lnt & sint \\ 2-t^2 & e^5t & t^{-2} \end{bmatrix}$$

$$= i \begin{bmatrix} lnt & sint \\ e^5t & t^{-2} \end{bmatrix} - j \begin{bmatrix} t-1 & sint \\ 2-t^2 & t^{-2} \end{bmatrix} + k \begin{bmatrix} t-1 & sint \\ 2-t^2 & t^{-2} \end{bmatrix}$$

$$= i(t^{-2}lnt - e^5 t sint) - j(t^{-2}(t-1) - sint(2-t^2))$$
$$+ k(e^5 t(t-1) - lnt(2-t)).$$

$F(t)$ is defined everywhere in the interval $(-\infty, \infty)$.

3. $F(t) = -3ti + (2t+1)j - (3t-2)k.$
The components are:

$$\begin{aligned} x &= x_0 + at = 0 - 3t. \\ y &= y_0 + bt = 1 + 2t. \\ z &= z_0 + ct = 2 - 3t. \end{aligned}$$

The point $(x_0, y_0, z_0) = (0, 1, 2)$ is on the line that is parallel to the vector $< a, b, c >=< -3, 2, -3 >$.

Exercise-7 on Page (82): Curves and Arc-Lengths.

1. $F(t) = (2t+1, 4t, 1-2t)$ then the length is:

$$L = \int_1^2 \sqrt{(f_1')^2 + (f_2')^2 + (f_3')^2} dt.$$

$$Where, f_1 = 2t + 1 \Rightarrow (f_1')^2 = 4.$$
$$f_2 = 4t \Rightarrow (f_2')^2 = 16.$$
$$f_3 = 1 - 2t \Rightarrow (f_3')^2 = 4.$$

$$Then, \ L = \int_1^2 \sqrt{24} dt = \sqrt{24}.$$

2. $F(t) = 3ti + 2\cos t j + 2\sin t k.$

$$L = \int_0^{\pi/2} \sqrt{(f_1')^2 + (f_2')^2 + (f_3')^2} dt.$$

$$Then, \ L = \int_1^2 \sqrt{9 + 4(\sin^2 t + \cos^2 t)} dt = \frac{\sqrt{13}}{2} t.$$

3. $F(t) = \cos 6ti + \sin tj + 4t^{3/2}t.$

$$L = \int_0^1 \sqrt{(f_1')^2 + (f_2')^2 + (f_3')^2} dt.$$

$$Then, \ L = \int_1^2 \sqrt{36 + 36t} \, dt = 9.$$

Curvature **K**, unit normal vector **N**, and the unit tangent vector **T** for the given functions:

4. $f(t) = ti + 2t^2 j - k.$

$$f(t) = \ <t, 2t, -t^2>.$$
$$f'(t) = \ <1, 4t, -3t^2>.$$

$$f''(t) = <0, 4, -6t>.$$

$$f' \times f'' = \begin{bmatrix} i & j & k \\ 1 & 4t & -3t^2 \\ 0 & 4 & -6t \end{bmatrix}$$

$$= 12t^2 i + 6tj + 4k.$$

$$And, \ |f' \times f''| = \sqrt{144t^4 + 36t^2 + 16}.$$

$$|f'|^2 = (1 + 16t^2 + 9t^4)^{3/2}.$$

$$Then, \ k = \frac{\sqrt{144t^4 + 36t^2 + 16}}{(1 + 16t^2 + 9t^4)^{3/2}}.$$

5. $f(t) = (1 + \frac{1}{\sqrt{2}}t)i + (1 - \frac{1}{\sqrt{2}}t)j.$

$$f = <(1 + \frac{1}{\sqrt{2}}t)i + (1 - \frac{1}{\sqrt{2}}t)j>.$$
$$f;(t) = <sqrt2, -1/sqrt2>.$$
$$f'' = 0, and,$$
$$k = 0.$$

6. $f(t) = 2ti + 1/tj$ at t=1.

$$f' = <2, -1/t^2>.$$
$$f'' = <0, 2/t^3>.$$
$$f' \times f'' = \begin{bmatrix} 2 & -1/t^2 \\ 0 & 2/t^3 \\ 0 & 4 \end{bmatrix} = 4/t^3.$$

At $t = 1$, then $k = 0.36$.

$$T = \frac{< 2, -t^2 >}{\sqrt{4 + 1/t}}.$$

$$T(1) = < \frac{2, -1 >}{\sqrt{5}}.$$

7. $f(t) = t^3 j - k \ at \ t = 0.$
$k = 0, \ T(0) = -1.$

Chapter-2
Exercise-1 on Page (86): Functions.

1. $x + y + z = 2$.
Let $z = f(x, y) = 2 - x - y$.
Then the graph of f is the set of all points (x, y, z) satisfying:
$z = 2 - x - y$ which is equation of the plane.
Its intercepts with $x - axis$ is the set: $y = z = 0 \Rightarrow x = 2$.
Its intercepts with $y - axis$ is the set: $x = z = 0 \Rightarrow y = 2$.
Its intercepts with $z - axis$ is the set: $y = x = 0 \Rightarrow z = 2$.
The trace of the plane on the coordinate planes are the lines connecting the intercepts as shown bellow:

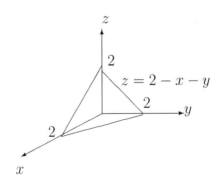

2. $3x - z = 1$.

$z = f(x) = 3x - 1$.

the graph of f is all points of 4x4 satisfying $z = 3x - 1$ Its intercepts with $x - axis$ is the set: $z = 0 \Rightarrow x = 1/3$.

Its intercepts with $z - axis$ is the set: $x = 0 \Rightarrow z = -1$.

Then the trace of the plane are the lines connecting intercepts.

3. $y = z^2$.

Here we can graph the function easier by making a table of points.

z	y
-2	4
-1	1
0	0
1	1
2	4

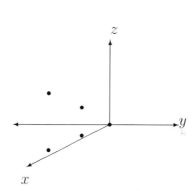

Which is a graph of a parabola opened to the left of $y - axix$.

4. $\frac{x^2}{4} + \frac{y^2}{9} = 16$.

Solving for y, gives:

$$y = \pm\sqrt{144 - \frac{9}{4}x^2}.$$

Which is equation of a circle with radius $r = 12$.

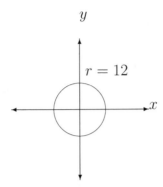

Exercise-2 on Page (96): Partial Derivative.

1.

$$f_x(x,y) = \frac{\partial}{\partial x}\{\frac{2x-y}{x+y}\} \Rightarrow y - fixed.$$

$$= \frac{3y}{(x+y)^2}.$$

2. $f_x = \frac{x}{\sqrt{x^2+y^2}}.$

3. $f_x = -1.$

4. $f_x = \frac{y}{2\sqrt{x}}.$

5. $f_x = \frac{2x}{x^2-y^2}.$

6. $f_x = 2xe^{x^2-y^2}.$

7. $f_x = e^x\{cos(xy) - ysin(xy)\}.$

Differentiation in limit form:

8. $f(x, y) = x - 3y.$

$$f_x = \frac{\partial f}{\partial x} = lim_{\triangle x \to 0} \frac{(x + \triangle x) - 3y) - (x - 3y)}{\triangle x} = 1.$$

$$And, \ f_y = \frac{\partial f}{\partial y} = lim_{\triangle y \to 0} \frac{x - 3(y + \triangle y) - x + 3y}{\triangle y} = -3.$$

9. $f(x, y) = x^2 + 2xy - y^2$. Using the same method we get:

$$f_x = 2x + 2y.$$
$$And, \ f_y = 2x - 2y.$$

10. $f(x, y) = \frac{4}{x+y}.$

$$f_x = f_y = \frac{-4}{(x + y)^2}.$$

slopes of surfaces in x, and y direction at the indicated point:

11. $f(x, y) = 5 - x^2 - y^2$, at $(2, 2, 3)$.

$$\frac{\partial f}{\partial x} = -2x = -4.$$
$$\frac{\partial f}{\partial x} = -2y = -4.$$

12. $f(x, y) = x^2 + y^2$, at $(3, 1, -2)$.

$$\frac{\partial f}{\partial x} = 2x = 6.$$
$$\frac{\partial f}{\partial x} = 2y = 2.$$

13. $f(x,y) = e^{-y}siny,\ at\ (1,0,0)$.

$$f_x = 1.$$
$$f_y = -1.$$

Chain Rule problems:

14. $z = x^2 + y$, $x = t^2 - 1$, and $y = 2 + t^2$. Then:
$z = (t^2 - 1)^2 + (2 + t^2)$.
Then, $\frac{dz}{dt} = 4t^3 - 2t$, and at $t = 1 \Rightarrow \frac{dz}{dt} = 2$.

15. $z = y\cos(xy)$, $x = \frac{1}{t}$, $y = 2t^3$, at $t = \pi/2$.

$$z = 2t^3\cos(2t^2).$$
$$Then, \frac{dz}{dt} = 2t^3(-sin(2t^2)(4t) + 6t^2\cos(2t^2)).$$
$$\frac{dz}{dt} = -8t^4 sin(2t^2) + 12t^5 \cos(2t^2).$$

16. $u = x^3 - 3x^2y + y^2$,, $x = s + 3t$, $y = 2s - t$.

$$u = (s+3t)^3 - 3(s+3t)^2(2s-t) + (2s-t)^2.$$
$$Then, \frac{du}{ds} = 3(s+3t)^2 - 6(s+3t)^2$$
$$-\ 6(s+3t)(2s-t) + 4(2s-t).$$
$$And, \frac{du}{dt} = 9(s+3t)^2 + 3(s+3t)^2$$
$$-\ 6(s+3t)(2s-t) - 2(2s-t).$$

17. $y = e^{y/x}$, $x = s + 2t$, $y = s - 2t$.

$$u = e^{\frac{s-2t}{s+2t}}.$$
$$Then, \frac{du}{ds} = e^{\frac{s-2t}{s+2t}} \cdot (\frac{s-2t}{s+2t})_s.$$

$$\frac{du}{ds} = e^{\frac{s-2t}{s+2t}} \cdot \frac{((s+2t)((1) - (s-2t)(1)}{(s+2t)^2}$$

$$= \frac{4t}{(s+2t)^2}.$$

$$And, \frac{du}{dt} = e^{\frac{s-2t}{s+2t}} \cdot \frac{((s+2t)((1) - (s-2t)(2)}{(s+2t)^2}$$

$$= \frac{-4s}{(s+2t)^2}.$$

18.$u = e^y cos y$, $y = s^2 - st - t^2$.

$$u = e^{s^2 - st - t^2}.$$

$$Then, \frac{du}{ds} = (2s-t)e^{s^2-st-t^2}\{cos(s^2 - st - t^@)$$

$$- sin(s^2 - st - t^2)\}.$$

$$And, \frac{du}{dt} = (s+2t)e^{s^2-st-t^2}\{sin(s^2$$

$$- st - t^@) + cos(s^2 - st - t^2)\}.$$

19. $z = \frac{x^2+y^2}{2xy}$, $x = 3t - s$, $y = t - 4s$.

$$Or, z = \frac{(3t-s)^2 + (t-4s)^2}{2(3t-s)(t-4s)}.$$

$$Then, \frac{\partial z}{\partial s} = \frac{-2xy(x+4y) + 10xy(x^2+y^2)}{4x^2y^2}.$$

20. $u = x^2 + y^2$, $x = scost$, $y = ssint$.

$$u = s^2cos^2t + s^2sin^2t.$$

$$then, \frac{\partial u}{\partial s} = 0.$$

$$And, \frac{\partial u}{\partial t} = 0.$$

Exercise-3 on Page (103): Gradient.

1. $f(x,y) = x^2 + y^3 + 2x - y^2$.
Then, the gradient of the function :
$\nabla f(x,y) = (2x + 2)i + (3y^2 - 2y)j$.

2. $f(x,y) = ycosx$.
$\nabla f(x,y) = -ysinxi + cosxj$.

3. $f(x,y) = 5x - 2y$.
Then, $\nabla f(x,y) = 5i - 2j$.

4. $f(x,y) = 10 - 3x + 5y^2$.
$\nabla f(x,y) = -3i + 10yj$, and at point $(2,1)$ is :
$\nabla f(x,y) = -3i + 10j$

5. $f(x,y) = cos(x + y)$.
then, $\nabla f(x,y) = -sin(x + y)i - sin(x + y)j$.
And, $\nabla(2,1) = -sin(2)i - sin(2)j$.

6. $f(x,y) = ln(x - y^2)$.
then, $\nabla f(x,y) = \frac{1}{x-y^2}i - \frac{2y}{x-y^2}j$.

And, $\nabla(3,2) = -i + 4j$.

Exercise-4 on Page (110): Lagrange-Multiplier.
1. The equation to be solved are:

$$
\begin{aligned}
1. \quad & V_x - \lambda g_x & = & \quad yz - \lambda(3y + z) = 0. \\
2. \quad & V_y - \lambda g_y & = & \quad xz - \lambda(3x + z) = 0. \\
3. \quad & V_z - \lambda g_z & = & \quad xy - \lambda(y + x) = 0. \\
4. \quad & g(x, y, z) & = & \quad 3xy + yz + xz - c = 0.
\end{aligned}
$$

Elimination lambda from the above equations we get:

$$
\begin{aligned}
1. \quad \Rightarrow \lambda & = \frac{yz}{3y + z}. \\
2. \quad \Rightarrow \lambda & = \frac{xz}{3x + z}. \\
3. \quad \Rightarrow \lambda & = \frac{xy}{y + x}.
\end{aligned}
$$

$$
\begin{aligned}
Then, \quad \frac{1}{\lambda} & = \frac{3y + z}{yz} = \frac{3x + z}{xz} \\
& = \frac{y + x}{xy}. \\
\Rightarrow \quad \frac{3}{z} + \frac{1}{y} & = \frac{3}{z} + \frac{1}{x} \\
& = \frac{1}{x} + \frac{1}{y}.
\end{aligned}
$$

Solving the last equalities gives : $\Rightarrow x = y = \dfrac{1}{3}z.$

2. Applying Lagrange-Multiplier as:

$$
\begin{aligned}
\nabla P(x, y, z) & = \lambda \nabla g(x, y, z). \\
And, \quad g(x, y, z) & = x^2 + 2y^2 + z^2 - 600 = 0.
\end{aligned}
$$

We start with : $\nabla P(x, y, z) = <2, 6, 4>$.
Then the equations to be solved are:

$$
1. \quad p_x - \lambda g_x = 2 - \lambda(2x) = 0.
$$

$$2. \quad p_x - \lambda g_y \quad = \quad 6 - \lambda(4y) = 0.$$
$$3. \quad p_x - \lambda g_z \quad = \quad 4 - \lambda(2z) = 0.$$
$$4. \quad g(x, y, z) \quad = \quad x^2 + 2y^2 + z^2 - 600 = 0.$$

Solving $(1 - 3)$ for λ gives:

$$1. \quad \Rightarrow \lambda \quad = \quad \frac{1}{x} = \frac{3}{2y} = \frac{2}{z}.$$

$$Or, \quad \frac{1}{\lambda} \quad = \quad x = \frac{2y}{3} = \frac{z}{2}.$$

Then the constrained equation will give:

$$x^2 + 2y^2 + z^2 \quad = \quad 600.$$
$$\frac{1}{\lambda^2} + 2(\frac{3}{2\lambda})^2 + \frac{4}{\lambda^2} \quad = \quad 600.$$
$$\frac{1}{\lambda^2}(\frac{19}{2}) \quad = \quad 600.$$
$$Then, \quad \lambda \quad = \quad \sqrt{\frac{19}{1200}} = 0.126.$$

Then: $x = \frac{1}{\lambda} = 7.95$, $y = \frac{3}{2\lambda} = 11.92$, and $z = \frac{2}{\lambda} = 15.89$.
And, the profit is:

$$P(x, y, z) \quad = \quad 2x + 6y + 4z.$$
$$= \quad 2(7.95) + 6(11.92) + 4(15.89).$$
$$P(x, y, z) \quad \Rightarrow \quad 150.98.$$

3. Surface area of the open-top box is:
$A(x, y, z) = xy + 2xz + 2yz$. Subject to the constrained
$xyz = c$. Applying the Lagrange-Multiplier Method:

$$\nabla A \quad = \quad \lambda \nabla g.$$
$$\nabla(xy + 2xz + 2yz) \quad = \quad \lambda \nabla(xyz - c).$$

The equations to be solved are:

$$
\begin{aligned}
&1. \quad A_x - \lambda g_x \quad \Rightarrow \quad y + 2z - \lambda(yz) = 0. \\
&2. \quad A_y - \lambda g_y \quad \Rightarrow \quad y + 2z - \lambda(xz) = 0. \\
&3. \quad A_z - \lambda g_z \quad \Rightarrow \quad 2x + 2y - \lambda(xy) = 0.
\end{aligned}
$$

Then solving for λ gives:

$$
\begin{aligned}
\lambda &= \frac{1}{x} = \frac{1}{y}. \\
\lambda &= \frac{1}{z} = \frac{2}{y}. \\
\lambda &= \frac{1}{z} = \frac{2}{x}.
\end{aligned}
$$

Or, $x = y = 2z$.
Using the constraint: $xyz = c$, gives: $x = y = (2c)^{1/3}$.

Chapter-3.

Exercise-1 on Page (119): Double-Integrals.

1. $\int_0^1 \int_1^2 (2x^2 + y^2 + xy + 4) dy dx$.

$$
\begin{aligned}
&\int_0^1 \int_1^2 (2x^2 + y^2 + xy + 4) dy dx \\
&= \int_0^1 [2x^2 dy + y^2 dy + xy dy + 4 dy]_1^2.
\end{aligned}
$$

$$= \int_0^1 (4x^2 + 8/3 + 2x + 8)dx.$$

$$\text{Then,} \quad \int_0^1 \int_1^2 (2x^2 + y^2 + xy + 4)dydx$$

$$= 4x^3/3 + x^2 + 32/3x]_0^1 = 13.$$

2. $\int_0^1 \int_0^1 x \, e^{x-y} \, dy \, dx.$

$$\int_0^1 \int_0^1 x \, e^{x-y} \, dy \, dx \quad = \quad \int_0^1 xe^x(\frac{-1}{e} + 1)dx.$$

$$= \quad (1 - \frac{1}{e}).$$

$$\text{Then,} \quad \int_0^1 \int_0^1 x \, e^{x-y} \, dy \, dx \quad = \quad 0.63.$$

Exercise-2 on Page (123): Iterated Integrals:

1. $\int_0^1 \int_0^\pi x \, siny \, dy \, dx.$

$$\int_0^1 \int_0^\pi x \, siny \, dy \, dx \quad = \quad \int_0^1 x[-cosy]_0^\pi.$$

$$= \quad \int_0^1 2dx = 2.$$

2. Find the volume of the region bounded on top by :
$z = 3x + y + 2$, on the bottom by the $xy - plane$, on the sides by the planes: $x = 0, x = 2, andy = 1, y = 2.$

$$V \quad = \quad \int \int f(x,y)dA.$$

$$= \int_0^2 \int_1^2 (3x + y + 2) dy dx.$$

$$= \int_0^2 (3x + 7/2) dx = 13.$$

Exercise-3 on Page (131): Multiple Integrals.

Use the triple integral to verify that the volume of a sphere of radius r is $\frac{4}{3}\pi r^3$.

It is easier to use polar coordinates here, where : $x = r\cos\theta$, and $y = r\sin\theta$, and equation of the circle is: $x^2 + y^2 + z^2 = a^2$, then in polar coordinates this equation becomes: $r^2 + z^2 = a^2$.

$$V = \int_0^{2\pi} \int_0^a f(r) r dr d\theta.$$

$$Since, \ x^2 + y^2 + z^2 = a^2.$$

$$Or \ z = f(r) = \sqrt{a^2 - r^2}.r dr d\theta.$$

$$Then, \ V = \int_0^{2\pi} \int_0^a f(r).r dr d\theta.$$

$$= \int_0^{2\pi} \frac{a^3}{3} d\theta.$$

$$then V = \frac{4}{3}\pi a^3.$$

$$Or, \ V = \frac{4}{3}\pi r^3.$$

Exercise-4 on Page (135): Triple Integrals.

1. Using the triple integrals , find the volume of the ellipsoid:

$$\frac{x^2}{4} + \frac{y^2}{4} + \frac{z^2}{16} = 1.$$

I_x : let $y = z = 0$, then $\rightarrow 4x^2 = 16$, $x = \pm 2$.
$I_y = f(x)$, let $z = 0 \rightarrow 4x^2 + 4y^2 = 16$, $y = \sqrt{4 - x^2}$.
$I_z = f(x, y) \rightarrow z = \sqrt{16 - 4x^2 - 4y^2}$. Then the intervals are:

$$V = 8 \int \int \int dV.$$

$$= 8 \int_0^2 \int_0^{\sqrt{4-x^2}} \int_0^{\sqrt{16-4x^2-4y^2}} dz dy dx.$$

$$= 8 \int_0^2 \int_0^{\sqrt{4-x^2}} \sqrt{16 - 4x^2 - 4y^2} dy dx.$$

$$V = \pi/4 \int_0^2 (4 - x^2) dx = \frac{64\pi}{3}.$$

Where, $\sqrt{16 - 4x^2 - 4y^2} = \sqrt{(4 - x^2) - y^2} = \sqrt{a^2 - y^2}$.

Then $\int \sqrt{a^2 - y^2} du = 1/2(u\sqrt{a^2 - y^2} + a^2 sin^{-1}(u/a))$.

2. Find the volume of the solid in the first octant bounded by the cylinder $y = x^2 + z^2 + 9$, the plane $x^2 + y = 4$, and the three coordinates.

I_x, let $z = 0 \rightarrow x = \pm 2$.
$I_z = f(x) \rightarrow z = 4 - x^2$.
$I_y = f(x, z) \rightarrow y = x^2 + z^2 + 9$. Then the intervals are:

$$\begin{aligned} I_x &= [0, 2]. \\ I_z &= [0, (4 - x^2)]. \\ I_y &= [0, (x^2 + z^2 + 9)]. \end{aligned}$$

Then the triple integral is:

$$V = \int\int\int dV.$$

$$= \int_0^2 \int_0^{4-x^2} \int_0^{x^2+z^2+9} dydzdx.$$

$$= \int_0^2 \int_0^{4-x^2} (x^2 + z^2 + 9)dzdx.$$

$$= \int_0^2 x^2z + 1/3z^3 + 9z]_0^{4-x^2} dx.$$

Then, V $\approx$ 72.

Chapter-4.

Exercise-1 on Page (148): Surface-Integrals.
2. Find the surface integral $\int\int_S(x + yz)\, dA$, where S is the portion of the plane $x + 2y + z = 2$ is in the first octant.

$$f_x = -1.$$
$$f_y = -2.$$
$$Then\sqrt{f_x^2 + f_y^2 + 1} = \sqrt{6}.$$
$$And, \int\int_S(x + zy)dA = \int\int_S[x + y(2 - x - 2y)]\sqrt{6}dA.$$
$$= \sqrt{6}\int_0^2 \int_0^{(1-x/2)} dydx$$
$$\int\int_S(x + zy)dA = \frac{14}{9}\sqrt{6}.$$

Exercise-2 on Page (154): Green's Theorem.
1. Use Greens theorem to evaluate $\int_c F.dr$, where C is a circle : $x^2 + y^2 = 4$, oriented counterclockwise, and $F = yi + 2xyj$.

$$\int_c f(x, y)dx + g(x, y)dy = \int\int(g_x - f_y)dA.$$

$$= \int\int_R (\frac{\partial N}{\partial x} - \frac{\partial M}{\partial y})dA.$$

$$= \int_0^2 \int_0^y (2y - 1)\, dx\, dy.$$

$$= \int_0^2 (2y^2 - y)dy.$$

$$= 2y^3/3 - y^2/2]_0^2.$$

$$\int_c f(x, y)dx + g(x, y)dy = \frac{10}{3}.$$

2. Use Greens Theorem to evaluate the integral: $\oint_c y^2 dx + x^3 dy$, where c is the path formed by the square with vertices on ; $(0, 0), (2, 0), (2, 2), (0, 2)$, on clockwise direction.

$$\int_c y^2 dx + x^3 dy = \int\int (g_x - f_y)dA.$$

$$= \int_0^2 \int_0^2 (3x^2 - 2y)dxdy.$$

$$= \int_0^2 (8 - 4y)dy. = 8y - 2y^2]_0^2.$$

$$\int_c y^2 dx + x^3 dy = 8.$$

Exercise-3 on Page (157): Divergence Theorem.

1. Use the divergence theorem to evaluate $\int\int_S F.nds$ where $F(x, y, z) = xi + yj - z^2k$, and S is the sphere : $x^2 + y^2 + z^2 = 4$.

$$
\begin{aligned}
f(x, y, z) &= xi + yj - z^2k. \\
\nabla.F &= 1 + 1 - 2z = 2(1 - z). \\
S : x^2 + y^2 + z^2 &= 4. \\
I_x &= [0, 2]. \\
I_y &= f(x) = [0, \sqrt{4 - x^2}. \\
I_z &= f(x, y) = [0, \sqrt{4 - x^2 + y^2}. \\
then, \int\int_S F.nds &= \int\int\int_R \nabla.F dV. \\
&= \int_0^2 \int_0^{\sqrt{4-x^2}} \int_0^{\sqrt{4-x^2-y^2}} dz\, dy\, dx.
\end{aligned}
$$

3. Compute $\int\int_S F.ds$ where $F(x, y, z) = -2xi + 3yj - 4zk$, bounded by the plane: $x + 2y + 2z = 4$.

$$
\begin{aligned}
F(x, y, z) &= -2xi + 3yj - 4zk. \\
\nabla.F &= -2 + 3 - 4 = -3. \\
then, \int\int_S F.nds &= \int\int\int_R \nabla.F(x, y, z)dV. \\
&= -3\int_0^4 \int_0^{2-x/2} \int_0^{2-x/2+y} dz dy dx. \\
\int\int_S F.nds &= 14.
\end{aligned}
$$

* * * * * *

4.13   Important Formula's

1. <u>Trigonometric Identities</u>

$$
\begin{aligned}
sin^2\theta + cos^2\theta &= 1. \\
1 + tan^2\theta &= sec^2\theta. \\
1 + cot^2\theta &= csc^2\theta. \\
sin(-\theta) &= -sin\theta. \\
cos(-\theta) &= cos\theta. \\
sin(2\theta) &= 2sin\theta\, cos\theta. \\
cos(2\theta) &= cos^2\theta - sin^2\theta. \\
&= 2cos^\theta - 1. \\
&= 1 - 2sin^2\theta. \\
cos^2\theta &= 1/2(1 + cos2\theta). \\
sin^2\theta &= 1/2(1 - cos2\theta).
\end{aligned}
$$

2. <u>Trigonometric Integrals</u>

$$
\int \frac{1}{x^2 + b^2}dx = \frac{1}{b}arctan(\frac{x}{b}). \ b \neq 0.
$$

$$
\int \frac{1}{x^2 - b^2}dx = \frac{1}{2b}log(\frac{x - b}{x + b}).
$$

$$
\int \frac{1}{b^2 - x^2}dx = \frac{1}{2b}log(\frac{x + b}{b - x}).
$$

$$
\int \frac{1}{\sqrt{x^2 + b^2}}dx = log(x + \sqrt{x^2 + b^2}).
$$

$$
\int \frac{1}{\sqrt{x^2 - b^2}}dx = log(x + \sqrt{x^2 - b^2}).
$$

$$\int \frac{1}{\sqrt{b^2 - x^2}} dx = arcsin\frac{x}{b}.$$

3. Integration by Parts

$$\int xe^x dx = xe^x - e^x.$$

$$\int logx dx = xlogx - x.$$

$$\int x\,logx\,dx = 1/2x^2\,logx - 1/4x^2.$$

$$\int x\,sinx\,dx = sinx - x\,cosx.$$

$$\int x\,cosx\,dx = cosx + x\,sinx.$$

4. Other Integrals

$$\int sin^2x\,dx = 1/2(x - sinxcosx).$$

$$\int cos^2xdx = 1/2(x + sinxcosx).$$

$$\int tan^2xdx = tanx - x.$$

$$\int sec^2xdx = 1/2\,secx\,tanx + 1/2\,ln\,|secx + tanx|.$$

$$\int \sin^2 x \cos^2 x\, dx \;=\; 1/8x - 1/32\sin 4x.$$

$$\int \sec^3 x\, dx \;=\; 1/2\sec x\tan x + 1/2\log(\sec x + \tan x).$$

$$\int \frac{\sqrt{x^2 - b^2}}{x}\, dx \;=\; \sqrt{x^2 - b^2} - b\sec^{-1}\left|\frac{x}{b}\right|.$$

$$\int \sqrt{b^2 - x^2}\, dx \;=\; \frac{x}{2}\sqrt{b^2 - x^2} + \frac{x^2}{2}\sin^{-1}\frac{x}{b}.$$

4. Exponential and logarithmic Integrals

$$\int x e^{bx}\, dx \;=\; \frac{1}{b^2}(bx - 1)e^{bx}.$$

$$\int (lnx)^n\, dx \;=\; x(lnx)^n - n\int (lnx)^{n-1}\, dx.$$

$$\int \frac{(lnx)^n}{x}\, dx \;=\; \frac{(lnx)^{n+1}}{n+1}.$$

$$* \quad * \quad * \quad * \quad * \quad * \quad * \quad * \quad * \quad *$$

4.14   Mathematica

In this section we will list some of the basic Mathematica calculations:

1. Dot-Product, cross-product, unit vector:

EX-1: Given 3-vectors:

$$u = \;<1, 1, -6>.$$
$$v = \;<3, 0, 8>.$$
$$w = \;-3, -1, 2>.$$

$Find:$

$a)$ $3v - 2w.$

$b)$ $2u - 3v + 4w.$

$c)$ $the\, length\, of\, u,\; or\; |u|$

$d)$ $the\, length\, of\, v,\; or\; |v|.$

$e)$ $A\, unit\, vector\, in\, direction\, of\, v.$

$f)$ $The\, angle\, between\, u\, and v.$

$g)$ $the\, cross\, product\, of\, u\, and\, v\, or\, area\, |u \times v|.$

h $Find\, the\, dot\, product\, of\, u\, and\, v$

$$in[1] := \;\{1, 1, -6\}$$
$$in[2] := \;\{3, 0, 8\}$$
$$in[3] := \;\{-3, -2, 2\}$$

$$a)in[4] := \;3v - 2w$$
$$out[4] = \;\{15, 2, 20\}$$

$$b)in[5] := \;2u - 3v + 4w$$
$$out[5] = \;\{-19, -6, -28\}$$

The Norm(Length) of a vector can be calculated in Mathematica using the "Norm" command:

$$c)in[6] : = Norm[u]$$
$$out[6] = \sqrt{38}$$

$$d)in[7] : = Norm[v]$$
$$out[7] = \sqrt{73}$$

Dot-Product:Use (.) such as "u.v" or the command "Dot[u,v]".
cross-Product: Use (x) or the command "Cross[u,v]".

EX: Find the equation of a line passing through two points: P_0= (1,2,3), and p_1 = (2, 4, 5), and graph the line.
Solution: line in Vector form is r(t) = $< x_0, y_0, z_0 > +$ t $< a, b, c >$.
Line in parametric form is: $x = x_0 + at, y = y_0 + bt, z + z_0 + ct$.
Clear[r,t];

$$r[t_-] = 1, 2, 3 + t2, 4, 5.$$

To graph the line: Use the command: Parametric-Plot3D.
To find the angle θ between two vectors, u, v use this command:

$$\theta = ArcCos[\frac{u.v}{Norm[u]Norm[v]}].Then, \ N[\%].$$

To find determinant of a matrix A use the command: Det[A].
To find a unit vector for u use the command:

Norm[u/Norm[u]].
To display in Matrix form use the command: Matrix-
Form[A]

Index